# MOUND-BUILDERS

AUSTRALIAN NATURAL HISTORY SERIES

# MOUND-BUILDERS

DARRYL JONES AND ANN GÖTH

CSIRO
PUBLISHING

National Library of Australia Cataloguing-in-Publication entry

Jones, Darryl N. (Darryl Noel)
Mound-builders: malleefowl, brush turkeys and scrubfowl/
Darryl Jones, Ann Göth

9780643093454 (pbk.)

Australian natural history series
Includes index.
Bibliography.

Megapodiidae – Ecology – Australia.
Megapodiidae – Habitat – Australia.
Megapodiidae – Behavior – Australia.
Megapodiidae – Nests – Australia.
Megapodiidae – Conservation – Australia.

Göth, Ann.

598.640994

Published by
**CSIRO** PUBLISHING
150 Oxford Street (PO Box 1139)
Collingwood VIC 3066
Australia

Telephone: +61 3 9662 7666
Local call: 1300 788 000 (Australia only)
Fax: +61 3 9662 7555
Email: publishing.sales@csiro.au
Web site: www.publish.csiro.au

*Front cover*
Malleefowl (photo by Jessica van Der Waag)

*Back cover*
Scrubfowl (photo by John Manger)

Set in 10.5/14 Adobe Palatino, Optima and Stone Sans
Edited by Lee K. Curtis, ataglance.com
Cover and text design by James Kelly
Typeset by Desktop Concepts Pty Ltd, Melbourne
Printed in China by Bookbuilders

# CONTENTS


| | | |
|---|---|---|
| Acknowledgments | | vii |
| 1 | Familiar yet distinct | 1 |
| 2 | Taxonomy, distribution and habitat | 7 |
| 3 | Appearance and ecology | 17 |
| 4 | The mound | 33 |
| 5 | Abandoned eggs | 57 |
| 6 | Growing up without parental care | 67 |
| 7 | Social and reproductive behaviour | 81 |
| 8 | Conservation and management of Australian mound-builders | 93 |
| Endnotes | | 107 |
| Index | | 117 |

# ACKNOWLEDGMENTS

No book is simply the product of its authors alone. For books such as ours, which attempt to survey a vast amount of published literature, the reality is that very little would have been possible without the efforts of a relatively small number of dedicated researchers. These extraordinarily motivated people, enduring the relentless heat of the mallee, the steaming jungles of the tropical north or the hazards of suburban backyards, have painstakingly changed our understanding and perceptions of these remarkable birds. They have also brought to light the precarious conservation status of certain populations, the complexities of managing recalcitrant urban megapodes and many other aspects about which we still know very little. To these magnificent megapoders – both past and present – we offer our sincere thanks.

We are especially grateful to belong to an international group of researchers and wildlife managers concerned with the conservation and study of the megapodes. The BirdLife/World Pheasant Association/IUCN Species Survival Committee Megapode Specialist Group provides a network for all those engaged in studying this family of birds. The willingness of its members to share information, views, suggestions and their unpublished data has been invaluable to us in bringing the latest findings to this book. In particular we would like to thank our colleagues, Joe Benshemesh, David Booth, Sharon Birks, Rene Dekker, Mark Hauber, David Priddell, Ross Sinclair and Jessica van der Waag, for their assistance and advice which they provided both directly and indirectly. We are also grateful to David Booth, Blair Parsons and Jessica van der Waag for permission to use unpublished data, and Heather Proctor and Jessica van der Waag for permission to reproduce their photographic images. We also thank Gwenda White who stepped in to provide drawings at the last minute.

Field work with megapodes can be time and energy consuming, and it is impossible to thank here all those field volunteers who helped us over the years. However, we would like to acknowledge the field support of Warren Canendo, Wayne Colbran, Ceinwen Edwards, Cici Legoe, Tom Nealson, Kirsty Nicols, Jeanette Nobes, Rachel Richardson, Geoff Ross, Uwe Vogel and Nicole Zimmermann. We also appreciate the support of all those many landholders who allowed us access to their land, for studying megapodes or their mounds.

Darryl Jones would also like to acknowledge the hospitality of Wayne Linklater and the staff of the Victoria University of Wellington, New Zealand, where much of the writing was completed, and Professor Stephen Davies for inviting him to attend the 2007 Malleefowl Forum in Katanning, Western Australia. Ann Göth's work on megapodes benefited greatly from the support of Darryl Jones, Chris Evans, and Mariella Herberstein, as well as the funds received through grants from Griffith and Macquarie University, the Dr. Otto Röhm Gedächnisstiftung, the Australian Geographic Society, and the World Pheasant Association.

We also express our great appreciation for the support provided by CSIRO Publishing, especially that of John Manger and Tracey Millen.

Finally, Darryl Jones sincerely thanks his family, Liz, Dylan, Caelyn and Manon, for putting up with the peculiarities of a preoccupied author. Ann Göth gratefully acknowledges the untiring support of her husband Wayne as well as the contributions of her newborn son Toby, who granted her the occasional quiet period for writing and allowed amazing insights into the difference between altricial and precocial young.

Darryl Jones
Brisbane

Ann Göth
Sydney

# 1
# FAMILIAR YET DISTINCT

*They all have one thing in common: they do not build a nest and brood their eggs, as do other birds, but bury them in the ground, or in a volcano, or in heaps of fermenting vegetable material built up for the purpose, and leave them to incubate and hatch alone. (H. Frith 1962: 1)*[1]

The small family of birds known as the 'mound-builders' have intrigued scientists and naturalists for almost 500 years. This interest is not due to spectacular plumage or melodious voices but because of their unexpected method of incubating their eggs. As long ago as 1521, Antonio Pigafetta, a member of Magellan's attempt to circumnavigate the world, encountered a fowl-like bird in the islands of the Philippines. In his memoirs[2] of the doomed voyage, Pigafetta recounted that the bird laid its very large ('duck-sized') eggs in holes in the sand where they were incubated through the heat of the sun. This description of incubation habits and egg size was quite accurate and, to our knowledge, the first mention of a mound-building species. Later accounts by other explorers were less accurate, often seeming to be exaggerated or simply fanciful: birds that produced eggs larger than themselves; egg pits that became furnaces; and chicks that could fly at hatching. This was typical of the sensationalism used by the authors of the times to pique interest in their travel books about exotic places. However, the combination of fact and fantasy that filled such accounts often rendered them simply too unreliable for serious scientific attention. Even two centuries later, when Buffon[3] was compiling the first comprehensive encyclopaedia of birds in 1786 (*Histoire naturelle des oiseaux*), he felt obliged to ignore these birds, the material available being 'fantastic and full of inconsistencies'.

While European naturalists may have been unconvinced about the existence of mound-builders, indigenous people throughout the islands of South-East Asia, the South Pacific and Australia had been aware of them for centuries. In virtually all of these locations, the eggs of the local mound-building species had been harvested, sometimes under carefully controlled conditions that ensured the sustainable production of eggs while preserving the adults. In some places, complex customs and lore developed that ensured the continuity of what was an invaluable source of protein.[4] In many other places, however, the overexploitation of eggs and hunting of adults quickly led to the extinction of species; this appears to have been the pattern that accompanied the steady spread of humans throughout the islands of the South Pacific.[5]

The formal 'discovery' of the mound-builders did not eventuate until the early 19th century, with the initial description of a new species collected by British naturalists from the colonies in New Holland (Australia). However, even this event was not straightforward: in 1821, when John Latham discovered the bird we now know to be the Australian Brush-turkey *Alectura lathami*, he seemed especially impressed by the naked red head and neck, and hence named the species the New Holland vulture.[6] This initial and superficial judgement was eventually rectified by the eminent ornithologist John Gould who later examined some specimens of a second species, collected by his Australian colleague John Gilbert in 1822. Gould correctly ascertained that this species – the Malleefowl *Leipoa ocellata* – was a distinct, yet definite member of the galliform group of birds, which also comprises the pheasants, quail and chicken-like species so abundant throughout Asia. Gould also noted the affinities with the first species – by then renamed the Australian Brush-turkey – and speculated that these two species would likely be separate components of this group of birds distributed widely through the Australian Oceanic region.[7]

In 1823, one year later, C. Dumont described the third Australian mound-builder, the Orange-footed Scrubfowl *Megapodius reinwardt*, which is found across the tropical north.[8] This species was, however, already known from areas outside Australia, as it has a much wider distribution than the other Australian mound-builders, and the first specimen had already been collected on the Indonesian island of Lombok.

The three Australian mound-builders each belong to separate genera and are representatives of three distinct developments within the family. Nonetheless, all exhibit the defining characteristic of their family: the construction of large mounds of decomposing organic material for the incubation of theirs eggs. Compared to the birds' body size, these massive mounds are by far the largest constructions made by any non-colonial animal (beaver dams, for example, are the work of groups of many animals working together).

At first glance, the mounds built by these birds may seem to be similar to the nesting mounds of crocodiles, which are piles of wet vegetation and mud. However, unlike mound-builders, crocodiles do not interact with the mound, apart from defending the site from potential predators. Mound-builders, in contrast, are often engaged in sophisticated manipulation of their mounds on a daily basis, adding new material, checking the internal temperature and adjusting the layers as required. Perhaps surprisingly – and unlike the crocodilians – these birds do not seek and gather up their emerging hatchings. Along with the use of mounds for incubation, a total absence of parental care is a defining characteristic of mound-builders.

All mound-builder chicks hatch underground, deep within the warm material that has provided the heat required for incubation. Other bird hatchlings usually hatch at the same time as the other young in the same clutch, but baby mound-builders hatch alone and independently. They must also dig their way unaided through a metre or so of dirt and twigs to reach the surface. This solitary beginning of life is inevitable for mound-builders because their eggs begin to develop immediately upon being laid into the warm substrate; most birds do not start to brood their eggs until the clutch is complete, ensuring that all of the chicks hatch at the same time. This enables the diligent parent to gather the brood together for warmth, protection and learning some crucial life skills. But not so the mound-builders.

With each egg developing from the time of laying, and a female laying many eggs at intervals of several days over a period of months, there is no feasible way a parent would be able to gather these chicks together. Instead, mound-builders produce the most precocial offspring of all birds: having spent hours or days determinedly digging their way to the top of the mound, the bundle of feathers that emerges is ready for life. It can run, find food, control its internal temperatures and (hopefully) evade predators. And on its first night out of the 'nest', it will roost alone in nearby vegetation.

## The Frith legacy

Although a number of studies of the Australian mound-builders had been undertaken early in the twentieth century, there is no question as to when and how our understanding of these birds came of age. In November 1950, a dedicated young scientist arrived at a small area of dense mallee bushland near the town of Griffith, in central New South Wales, with the intention of studying an undisturbed and relatively abundant population of Malleefowl in a nature reserve surrounded by vast expanses of wheat. H. J. ('Harry') Frith had previously attempted to study Australian Brush-turkeys in the subtropical rainforests of northern New South Wales, but these plans had been thwarted by the relentless hunting of this large 'game' bird. He thus looked for opportunities elsewhere and was not deterred by the fact that the dry and harsh semi-arid mallee country appeared to be an entirely unsuitable environment for a mound-builder to make a living. It was here that Frith slowly and laboriously began his work on Malleefowl that would completely reconstruct our knowledge of this species and the family to which it belongs. Not only was Harry Frith an exceptional scientist, he was also a gifted and determined communicator, bringing his many discoveries to a worldwide audience through magazine articles, scientific papers and appearances in the print and broadcasting media of the day. Finally, his efforts were summarised in one of the first Australian examples of a highly successful popular science book, *The Mallee-Fowl: The Bird that Builds an Incubator*, published in 1962.

In some ways, Frith was almost too successful in promoting his groundbreaking studies of the Malleefowl around the world. Even today, ornithologists and birdwatchers are likely to cite the Malleefowl as *the* mound-builder; due to Frith's efforts, it is still the most well-known. The Malleefowl is, however, the most atypical of the entire family. This species is the only one to live in a dry environment; all other species are usually found in dense, moist forests throughout the tropics and subtropics. Mound-builders almost certainly evolved in the steaming jungles of New Guinea, where the construction of a mound of decomposing organic matter would have been relatively straightforward. In comparison, the effort required to construct and maintain a mound in the mallee seems almost impossibly difficult.

Today, although a vast amount of research into mound-builders has been conducted since Frith's time, it remains impossible to underestimate the legacy and influence of this scientist. In many ways, all subsequent work is built on the Frith foundation, and for this reason we are pleased to acknowledge this pioneer of Australian ornithology through quotations from his works. Although very different in character and aim to Frith's book, we hope that this small volume may be regarded as a suitable means of describing progress since.

## Mound-builders are megapodes

The mound-builders belong to the family of birds known as the Megapodiidae or the 'megapodes'. This name – *mega-pod* – refers to the exceptionally large feet and toes of all the species, an obvious adaptation to the raking of leaf litter and digging in the substrate required for the construction of incubation mounds. However, while most of the 22 species within the family Megapodiidae do indeed build mounds, a number of species use other external heat sources for incubation. While the mound-builders use the heat source produced by micro-organisms engaged in the decomposition of damp organic matter, several other species have discovered that the soil at specific depths in geothermal areas, such as occur in locations throughout south-east Asian and the south Pacific, are entirely suitable for incubation. Others have learned to lay their eggs in shallow holes on beaches of dark sand that have been heated by the sun. The species that do not construct mounds, but utilise heated soil and sand below the surface, are often referred to as 'burrow-nesters' to distinguish them from the 'mound-builders'. However, the three Australian species all use mounds, and in this book, we hence refer to them collectively as 'the mound-builders', while we use the term 'megapodes' for the entire family.

For simplicity, we generally refer to the three Australian mound-builders by the simplified names of Brush-turkey (assuming the 'Australian'), Malleefowl and Scrubfowl (assuming the 'Orange-footed'), unless we are referring to non-Australian species.

# 2

# TAXONOMY, DISTRIBUTION AND HABITAT

> *On the mainland of Australia and the islands to the north lives a family of rather dull-looking black or brown birds about the size of domestic fowl. Called megapodes after their big feet, they rarely fly, have raucous calls and are seldom seen by man. They are of interest because they do not brood their eggs as other birds do. (H. Frith 1959: 52)*[1]

Burying eggs in piles of warm soil or sand is a rather unusual practice for a bird. In fact, it is very similar to incubation in crocodiles and some turtles, who either gather together large piles of rotting vegetation on the banks of rivers, or burrow into sun-exposed sand. Indeed, megapodes have a number of other reptilian features including the relatively low incubation temperatures, the long development time of the eggs before hatching, the high proportion of yolk in the egg and the well-developed state of the young upon hatching. These apparently primitive features lead some naturalists to suggest that megapodes must be directly descendent from their reptilian ancestors, possibly even representing the most primitive living birds.[2] However, this was not a view that persisted for long: careful examinations of their morphology made it clear that megapodes were normal modern birds, which had developed their incubation practices later in their evolution.

Although we are now certain that megapodes belong to the Galliformes – a large group of birds especially abundant in the Asian region – early attempts to determine taxonomic relationships suggested links with waders, pigeons, lyrebirds, and even vultures (recall the Australian Brush-turkey initially being called the 'New Holland Vulture').[3] However, by about the 1840s most authorities were convinced that the megapodes were typical galliforms, and one of the many members of the group that included the quail, pheasants, grouse and chickens.

The attention of taxonomists then turned to trying to understand the relationships within this group and within the megapode family – an exercise that has continued with considerable debate and disagreement ever since.[3,4] Whereas taxonomists have traditionally based their complex arguments on detailed comparisons of skeletons, plumage, physiology, endocrinology and even ectoparasites, recent approaches have included a range of powerful new methods and sources of information simply unavailable to earlier taxonomists. In particular, techniques that compare the genetic similarities and use molecular features to reconstruct evolutionary histories are profoundly changing our understanding of the taxonomic relationships between groups of birds. The latest study[4] to determine the relationships of the megapode family (the Megapodiidae) with other birds places the family as one side of two major branches of the evolutionary tree, with the other branch containing all the other galliforms (Figure 2.1).

## Megapode distribution: dispersion and extinction

For a relatively small family of birds, the megapodes are found over a remarkably broad geographical range in the Indo-Pacific region, being spread across a large part of the world's most fragmented and dispersed pieces of land (Figure 2.2). At least nine species of megapode occur on the large New Guinea landmass (the island that includes Papua New Guinea and West Papua) or islands immediately adjacent (such as Waigeo, Biak and New Britain), and it is likely that this region was the group's original centre of evolution.

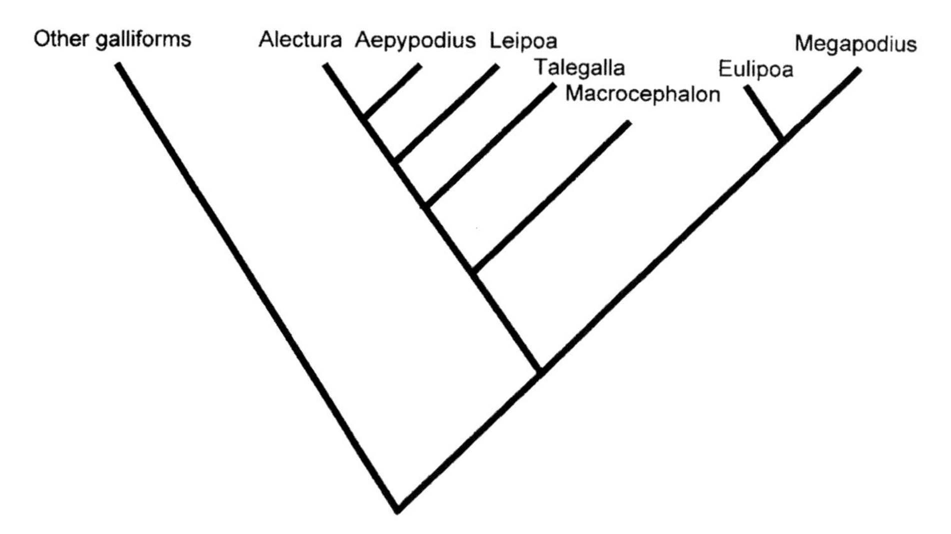


**Figure 2.1 Possible evolutionary relationships between the megapode genera.**
Adapted from Birks & Edwards (2002).

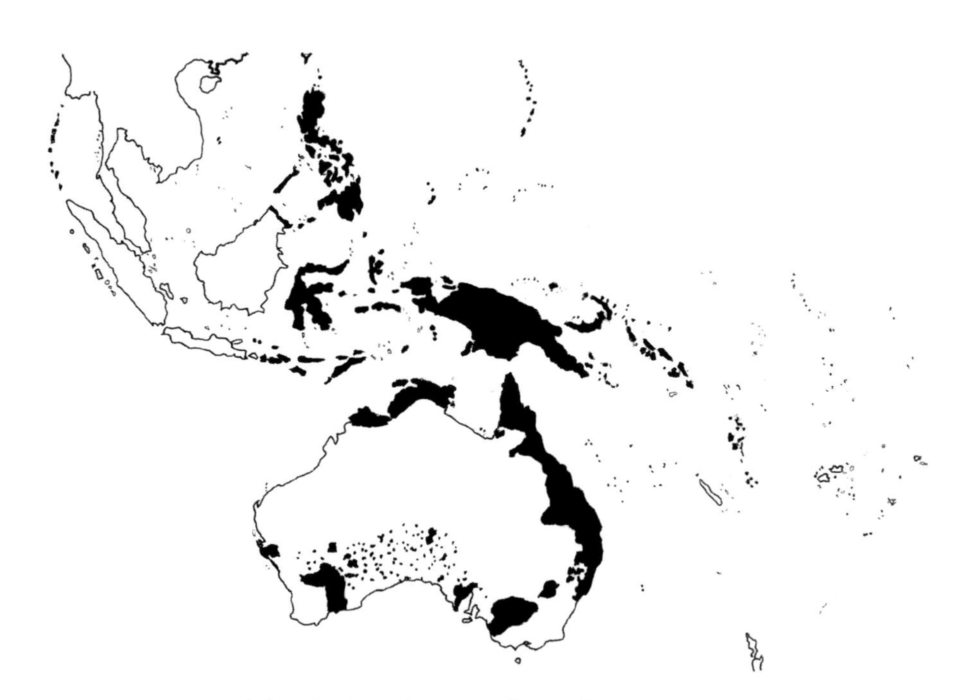

**Figure 2.2 Regional distribution of megapode species.** Drawing: Gwenda White

In general, most of the megapode genera are confined to relatively contiguous geographical ranges, for example *Macrocephalon* to Sulawesi, *Tallegalla* and *Aepypodius* to New Guinea, and *Leipoa* to inland Australia. The exception to this pattern is the genus *Megapodius*, the scrubfowl. Despite their small size, representatives of this group have spread out in all directions from their origins around present-day eastern New Guinea, eventually occurring west to the Nicobar Islands in the Indian Ocean (93°E), east to the island of Niuafo'ou (175°W) in the Tongan archipelago, north to the Mariana Islands (18°N) and south to the Kermadec Islands (29°S).

## Megapodes: genus and species, mound and burrow

After considerable discussion, most taxonomists now accept that the family of megapodes consists of seven genera with 22 extant species (Table 2.1 summarises these genera and lists the species within each). Bear in mind, however, that an additional 30 species have become extinct over the last 1000 years, almost all because of human activities in the islands of the South Pacific. The evolutionary relationships between these genera (their 'phylogeny') have been much debated and many arrangements suggested. The phylogeny provided here[4] (Figure 2.1) is the most recent and is based on analysing nuclear and mitochondrial DNA samples from all genera and 15 species. In short, this phylogeny largely corresponds with earlier ones that were based on the birds' physical features of feathers, nostrils, oil glands, feet and eggs.[5] This arrangement separates the family into the exclusively mound-building genera and those genera that may also incubate their eggs in burrows ('burrow-nesters'). Some exceptions occur, such as the Maleo *Macrocephalon maleo* from Sulawesi, which is one of the few megapodes that does not construct a mound even though, according to the DNA analysis, it is nevertheless closely related to the mound builders.

The details of such discussions about phylogenetic relationships among other megapode species are rather peripheral to this book. However, the phylogeny provides two important insights into the relationship among the Australian species. First, the megapode family is monophyletic: all species, including the Australian ones, have a common ancestor. Second, the three Australian megapodes represent quite distinct strands of evolution within the family as a whole, despite occurring on the same continent.

The rest of this book explores the similarities and differences between these very divergent species in more detail. In the following section, we first describe some extinct species of Australian megapodes and then

Table 2.1 Megapode genera and species (Australian species in bold)

| General megapode group | Genus | Species |
|---|---|---|
| The brush-turkeys | *Alectura* | *Alectura lathami*<br>**Australian Brush-turkey** |
| | *Aepypodius* | *Aepypodius arfakianus*<br>Wattled Brush-turkey |
| | | *Aepypodius bruijnii*<br>Bruijn's Brush-turkey |
| The talegallas | *Talegalla* | *Talegalla cuvieri*<br>Red-billed Talegalla |
| | | *Talegalla fuscirostris*<br>Black-billed Talegalla |
| | | *Telegalla jobiensis*<br>Brown-collared Talegalla |
| The malleefowl | *Leipoa* | *Leipoa ocellata*<br>**Malleefowl** |
| The maleo | *Macrocephalon* | *Macrocephalon maleo*<br>Maleo |
| The scrubfowls | *Euliopoa* | *Eulipoa wallacei*<br>Moluccan Scrubfowl |
| | *Megapodius* | *Megapodius pritchardii*<br>Polynesian Scrubfowl |
| | | *Megapodius laperouse*<br>Micronesian Scrubfowl |
| | | *Megapodius nicobariensis*<br>Nicobar Scrubfowl |
| | | *Megapodius cumingii*<br>Philippine Scrubfowl |
| | | *Megapodius bernsteinii*<br>Sula Scrubfowl |
| | | *Megapodius tenimberensis*<br>Tanimbar Scrubfowl |
| | | *Megapodius freycinet*<br>Dusky Scrubfowl |
| | | *Megapodius geelvinkianus*<br>Biak Scrubfowl |
| | | *Megapodius forstenii*<br>Forsten's Scrubfowl |
| | | *Megapodius eremita*<br>Melanesian Scrubfowl |
| | | *Megapodius layardii*<br>Vanuatu Scrubfowl |
| | | *Megapodius decollatus*<br>New Guinea Scrubfowl |
| | | *Megapodius reinwardt*<br>**Orange-footed Scrubfowl** |

discuss, for each of the three species occurring in Australia today, its place within the evolutionary history of the megapodes, as well as their present-day distribution.

## Extinct Australian megapodes

So far, two extinct species of megapodes have been discovered as fossil deposits. The first of these, *Progura gallinacea,* is a species the skeletal remains of which have mainly been found in a broad area of high country inland regions, from the Bluff Downs in north-west Queensland to near Naracoorte in South Australia. It was originally described as two species because of a marked difference in size. It is now suggested that this was due to the fact that males were much larger (5–7 kg) than females (4–5 kg), and that all the fossils belong to a single species. Compared to the largest modern-day species (for example, the Australia Brush-turkey weighs about two kilograms) this species was considerably larger, hence its common name of Australian Giant Megapode.[6] This species has been extinct for about 34 000 years, its demise attributed primarily to the changing climatic conditions of inland Australia. Nonetheless, its obvious morphological similarities to living megapodes suggest it would also have constructed mounds, which may have been substantial features of the landscape. But even these would dwarf in comparison to the real giant among megapodes, *Sylviornis neocaledoniae,* a flightless monster from New Caledonia, which weighed 30–40 kg. This species constructed massive mounds up to 50 m across and 5 m high, which are still evident today, centuries after it disappeared.[7]

The second fossil megapode from Australia represents the very other end of the size spectrum. *Ngawupodius minya,* recently described from deposits in Lake Pinpa, South Australia, was only about 230–330 g in weight – about two-thirds the size of the smallest living species.[8] Moreover, it is also the oldest fossil megapode to have been discovered, occurring during the late Oligocene (34–24 million years ago). This is an important discovery as it places the family firmly within the Australian landmass, at a considerably earlier period than previously suspected.

## Australian megapodes today

The three extant Australian megapodes are found over a large part of the continent. Their distributions are mainly non-overlapping (Figure 2.3), an indication of their adaptation to rather different habitats.

Figure 2.3 Distributions of the three Australian mound-builders: Australian Brush-turkey (diagonal lines); Malleefowl (solid black); and Orange-footed Scrubfowl (horizontal lines). Drawing: Gwenda White

## *Australian Brush-turkey*

The Australian Brush-turkey is one of three species within the 'brush-turkey' group, and the only representative from Australia. The name 'brush' refers to the species's preference for rainforest habitats, or 'brush' in the terminology of the early European naturalists, while the 'turkey' term was previously often used for large, ground-dwelling birds that were regarded as potential New World game species (such as 'Plains Turkey' for the Australian Bustard *Ardeotis australia* of the inland savannahs).

All three species of this group are large, black-plumaged birds which construct mounds on the floor of dense forests of either eastern Australia (the Australian Brush-turkey), central New Guinea (the Wattled Brush-turkey) and the island of Waigeo (Bruijn's Brush-turkey), a large island to the north-west of the New Guinea. These birds are the only megapodes that wear a bizarre array of brightly coloured and inflatable neck wattles and head combs, along with strongly pigmented bare facial and neck skin.

Not unexpectedly, these adornments appear to be employed in mate attraction, with the brush-turkeys being the only megapodes with a distinctly non-monogamous mating system.

Until recently, very little was known about the two non-Australian species, with the only extensive study of the Wattled Brush-turkey being conducted on captive birds in Germany.[9] The third species was known primarily as one of the world's rarest birds, with the last confirmed sighting being made in 1938. Despite numerous attempts to relocate the bird, nothing was forthcoming until 2002 when Iwein Mauro, a determined young Belgian ornithologist, made the first recent observations of this species in the wild and obtained remarkably detailed information on many aspects of its ecology and behaviour.[10]

The Australian Brush-turkey is found on a north–south axis along the moist forests of the mountains and coasts of eastern Australia, from Cape York to the Illawarra area south of Sydney, New South Wales. It is typically associated with closed forests including rainforests, wet sclerophyll and even mangrove forests. However, it is also remarkably flexible and can live successfully in a wide range of other habitats. An example of this was the species's apparent expansion of range during the infamous infestation of the noxious Prickly Pear *Opuntia* spp.[11] This virulent cactus covered 25 millions of hectares of New South Wales and Queensland. Although reliable information remains patchy, the distribution of Brush-turkeys apparently expanded to cover most of the plant's range. When the cactus was finally brought under control by the celebrated *Cactoblastis* moth, the Brush-turkey returned to its former range, although some populations may have been left isolated far to the west. This has been suggested as a possible explanation for the existence of the species in remote areas such as the Pilliga (now apparently extinct) in northern New South Wales and Carnarvon Gorge in central Queensland.

Apart from the expansion related to the spread of Prickly Pear, the Brush-turkey seems to have contracted significantly from its original range during much of the 20th century.[12] This appears to be largely the result of wide scale habitat destruction along the east coast, and the continuous hunting throughout its distribution. Chapter 8 will discuss this in more detail, but it is worth noting that there are reliable historical records of the species as far south as Cape Howe (37.5°S) and near Jindabyne (36.4°S).[13]

### *Malleefowl*

For years, scientists speculated how this arid-dwelling species fits in with the other megapodes, all of which inhabit tropical jungles or at least densely vegetated areas. The recent phylogeny (Figure 2.1) provides strong evidence that the Malleefowl – the sole member of its genus – is more closely related to the brush-turkeys than to other groups within the family. This supports the proposition that the ancestors of this species were 'typical' forest megapodes, occupying the interior of the Australian continent when the climate was significantly wetter than at present. Ancient Malleefowl presumably constructed their mounds in the denser vegetation of the time but were able to adapt to the relentless drying that occurred over the past several thousand years. Although this is unquestionably one of the great stories of survival, the precarious status of the species today suggests that the species's millennia of adaptations may now be close to its limits.

Clearly, the Malleefowl has not responded well to the continuing destruction and degradation of its habitat, competition for food and the impact of introduced predators. Originally, the Malleefowl was found throughout many parts of the semi-arid (receiving 200–450 mm of rain annually) lands of southern Australia, from central Western Australia to central New South Wales, but also into the very centre of the continent. From this vast area, the current range of the species continues to contract and fragment. Today, Malleefowl appear to have been forced to occupy wooded remnants in lower rainfall (220–300 mm) areas, which is often sub-optimal non-mallee country. Chapter 8 discusses the Malleefowl's demise in more detail.

### *Orange-footed Scrubfowl*

The Orange-footed Scrubfowl belongs to the largest and most widely distributed group within the megapode family, the genus *Megapodius* or scrubfowl.

The 14 species of scrubfowl are comparatively small and rather dull coloured birds, with little decoration beyond a moderate crest of feathers and pigmented facial skin in some species. All scrubfowl construct mounds, except for two species which are exclusively burrow-nesters (the Moluccan Scrubfowl *Megapodius wallacei* and Polynesian Scrubfowl *Megapodius pritchardii*), and except for at least four species that occasionally also lay their eggs in burrows excavated into geothermal areas or sun-heated beaches.

The Orange-footed Scrubfowl has a thoroughly tropical distribution and occurs across most of the entire northern edges of the continent, with the only gap in this continuous distribution occurring across the Gulf of Carpentaria. It is found primarily in coastal monsoon rainforests, mangroves and forested off-shore islands, although it also occurs in higher altitudes in the cooler rainforests of the Atherton Tablelands of northern Queensland.

It has been the least impacted of the Australian species and appears to occupy most of its historic range within the continent. There is, however, some suggestion of a reduction in its southern range, along the coasts of northern Queensland, and of a disappearance from the savannahs of northern Australia during pre-European times. Chapter 8 describes the potential reasons for such changes in its range in more detail.

# 3

# APPEARANCE AND ECOLOGY

*The megapodes affinities are undoubtedly with the gallinaceous or fowl-like birds, and their most obvious physical characteristic is the proportional great size and power of the feet and legs. This peculiarity is correlated with the mound building habit, for it is with these implements that the birds gather the huge heaps of debris which serve as incubators for the eggs. (C. Barrett 1931: 107)*[1]

As we have explained, the three mound-builders living in Australia today represent three rather different strands of megapode evolution which inhabit rather different habitats. One is typical of tropical jungles, one is a generalist rainforest edge dweller and, most unexpectedly, one makes its home in the dry arid interior. Obviously, adapting to such places will have led to many differences in behaviour and activities. In this chapter we describe some of the general features of the three Australian mound-builders and explore their similarities and differences. We focus primarily on their appearance, vocalisations, feeding habits and basic ecology and behaviour. Details on reproduction, incubation and conservation are provided in other chapters.

## Appearance and vocalisations

The relatively large size and dark colours of mound-builders apparent from identification books may suggest that they are easily detected – even conspicuous – in the wild. In reality, however, each species is remarkably difficult to spot and all are exceptionally well camouflaged in their preferred habitat.

### *Australian Brush-turkey*

*Appearance* In the forests and thickets where they usually build their mounds, Brush-turkeys blend in with their environment so well that most people are unlikely to be aware of them. With their predominantly black or brownish plumage, Brush-turkeys are hard to make out against the deep shade of the background vegetation. In addition, in undisturbed areas they are remarkably shy birds, actively avoiding people. They are also primarily solitary, making them even harder to observe. The best method to detect the presence of Brush-turkeys is to listen for the noise these birds make while scratching for food in the leaf litter or while building their incubation mounds.

Despite the difficulties in seeing Brush-turkeys in undisturbed environments, many people have encountered them in picnic areas, camping grounds and increasingly in suburban gardens, especially along coastal Queensland and New South Wales. Having previously been relatively rare and extremely difficult to approach – presumably because they were actively hunted – Brush-turkeys have adapted remarkably well to human-dominated environments, particularly in places were they can scavenge for food scraps. Here, out in the open, they often congregate in groups and their main physical features become more obvious: the black and grey-brown plumage, the brightly coloured head, vertically fanned tail and large size, approximately equal to that of a domestic hen turkey.

The most conspicuous part of the bird's appearance is the bare deep red skin of the adult's head and neck, with a section of yellow or pink loose skin separating the red neck from the black feathers of the body. In most females and subadults, a few black bristles cover the red colour on the head, while in males and some females, such bristles are sparse and the head appears bright red. Outside the breeding season (typically the first half of the year), the colour of the head appears duller.

The loose skin section of the lower neck is known as the 'wattle'. Here the skin extends into a fleshy wrinkled fold that varies in colour from dull to bright yellow. The sub-species of Brush-turkey which occurs in northern

Cape York (*Alectura lathami purpureicollis*) has a mauve, pinkish to purple-white wattle instead of the yellow, but is otherwise identical to the main subspecies (*Alectura lathami lathami*). The size of the wattle differs enormously between the sexes, and also during different times of the year. When not breeding, the wattle of both females and males is typically a loose area of pale bare flesh, and the sexes are almost indistinguishable. During the breeding season, the influence of reproductive hormones leads to major changes in the appearances of the sexes. In females, the wattle becomes looser and slightly folded, but does not enlarge greatly. In breeding males, however, it expands into an elongated loose and pendulous pouch sometimes long enough to reach the ground.

Males are able to control the size of their wattles and it may be retracted or extended quickly. We have caught many adult males with impressively large wattles only to find these quickly retracted into a small ruffle resembling a cravat. It is not yet known how males influence the size of their wattle although it is likely to involve the pumping of either air or blood into the fleshy structure. In the wild, males appear to use their wattles during social interactions as highly conspicuous social signals. Our observations indicate that socially dominant males fully extend their wattles while subordinates retract their wattles in the presence of more dominant individuals. Wattles are also inflated to produce the 'boom' vocalisations (see pages 21–23).

An additional role for the wattle was recently suggested when it was discovered that the pale skin reflected ultraviolet light brightly.[2] It is now well established that birds see far more of the light spectrum than humans, including well into the UV component. This has led to significant rethinking of our understanding of the visual world of birds, and considerable study is underway that explores the UV reflectance plumage and soft parts of many birds. The fact that Brush-turkey wattles 'glow' in the dim world of the rainforest floor may provide another means of visual communication for these birds. However, this aspect requires more detailed study.

While the possession of a larger wattle helps distinguish males from females in the breeding season, at other times the sexes can usually be separated by general differences in their size. Males are larger, which is most obvious when comparing the length of their legs (tarsii). On average, the male tarsus is 10–12 mm longer than the female tarsus (Table 3.1), although this difference is less obvious in the field. The larger size of males is usually reflected in their greater weight, but females may weigh as much as males when they are carrying eggs within (Table 3.1).

**Table 3.1 Tarsus length and body mass of adult male and female Australian Brush-turkeys from New South Wales (NSW) and Queensland (QLD).**

| | Tarsus | | | | Body mass | | | |
|---|---|---|---|---|---|---|---|---|
| | **Mean** | **S.D.** | **Range** | **n** | **Mean** | **S.D.** | **Range** | **n** |
| NSW females | 95 | 4.4 | 86.9–105.7 | 17 | 2227 | 244 | 1850–2670 | 17 |
| NSW males | 104.8 | 3.1 | 98.3–108.9 | 16 | 2441 | 168 | 2150–2750 | 16 |
| QLD 1 female* | 104.2 | 4.5 | 98.4–112.6 | 12 | 2103 | 197 | 1840–2430 | 12 |
| QLD 1 male | 116.5 | 8.1 | 103.5–126 | 16 | 2579 | 200 | 2310–2950 | 16 |
| QLD 2 females* | 90.1 | 4.71 | 84.0–99.2 | 12 | 2160 | 250 | 1980–2510 | 23 |
| QLD 2 males | 100.9 | 3.21 | 97.0–106.0 | 12 | 2520 | 180 | 2120–2950 | 37 |

*a direct comparison of tarsus length in the different populations is not possible because different methods of measurement were used. S.D. = Standard deviation.
Source of data: NSW: one population (Pearl Beach) measured by Göth, Ross and Nicols in 2004–2005, QLD 1: one population (Maleny) measured by Jones and Göth in 1998–1999; QLD 2: museum skins from throughout Queensland, reproduced from Jones *et al.* (1995) with permission.

With a more detailed examination (requiring capture of the birds), the sexes can be separated by examining the cloaca, a technique we found to be useful even among very young birds. Both sexes have a phallic structure on the ventral lip of the cloaca, but in males this is considerably larger and normally darker pink in colour (see Figure 3.1). Interestingly, no other megapode appears to possess these structures, although only the Malleefowl has been carefully examined; their presence is a valuable and straight-forward method of sexing even very young Brush-turkeys.

In comparison with the other two Australian mound-builders, Brush-turkeys are not only larger, but also the only species with a naked head, a wattle and a long conspicuous tail that is vertically fanned. This orientation of the long tail feathers appears to aid the species when running and manoeuvring rapidly through dense vegetation – a valuable adaptation when being pursued by a predator or dominant bird. Sometimes, the birds spread their tail feathers out widely, while at other times they fold their tail into a much narrower extension. It appears that this spreading or folding of the tail is associated with social interactions with other Brush-turkeys, and that dominant birds generally spread their tail more widely than do subordinates. However, more behavioural observations are required to confirm these assumptions.

Tail feathers of such length do, however, pose one substantial risk: a predator, such as a quoll or dingo, could catch the bird by grabbing the tail. As a result, Brush-turkeys have developed the ability to shed all their tail feathers almost instantly, much like a lizard loses its tail.

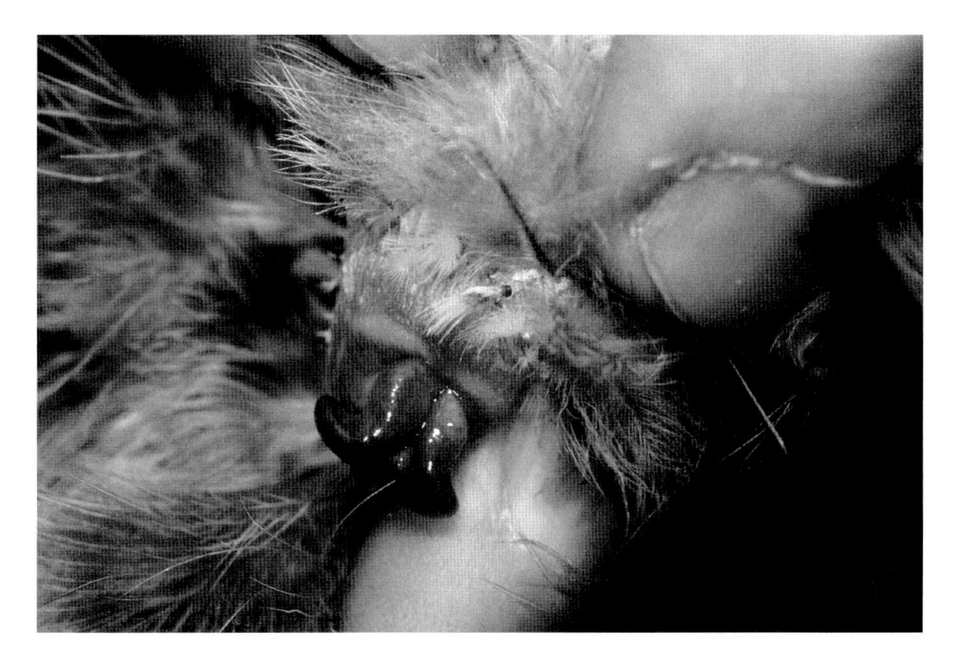

**Figure 3.1 Cloacal structures of adult male Australian Brush-turkey.** Photo: Darryl Jones

***Vocalisations*** Of the three Australian megapodes, only the Scrubfowl can be described as being particularly vocal (see pages 25–27). The calls of the Brush-turkey are limited to a series of unsophisticated grunts and low frequency noises, primarily used for communication between individuals over short distances, or apparently when a bird is frightened or distressed.

Four specific calls have been recognised.[3] The main contact call is a very deep, low-pitched and repeated gulping sound generally used when males and females are near one another, especially in the vicinity of the mound. Usually only a few notes are produced. The second call is rather similar to the contact call but is not as deep and is frequently more rapid and repetitive; this seems to be typically associated with some level of disturbance. The third call is a loud, crowing noise produced with the bill wide open. This was commonly produced when we were handling birds, and was heard once when a dingo caught a female. Presumably this vocalisation is limited to the extreme distress associated with predation.

An unusual fourth mode of noise production in Brush-turkeys is the 'boom', which is produced by forcing air from the inflated wattle through the nasal openings. As only males develop a larger wattle around the base of the neck, they are the only sex to produce this sound. In order to boom,

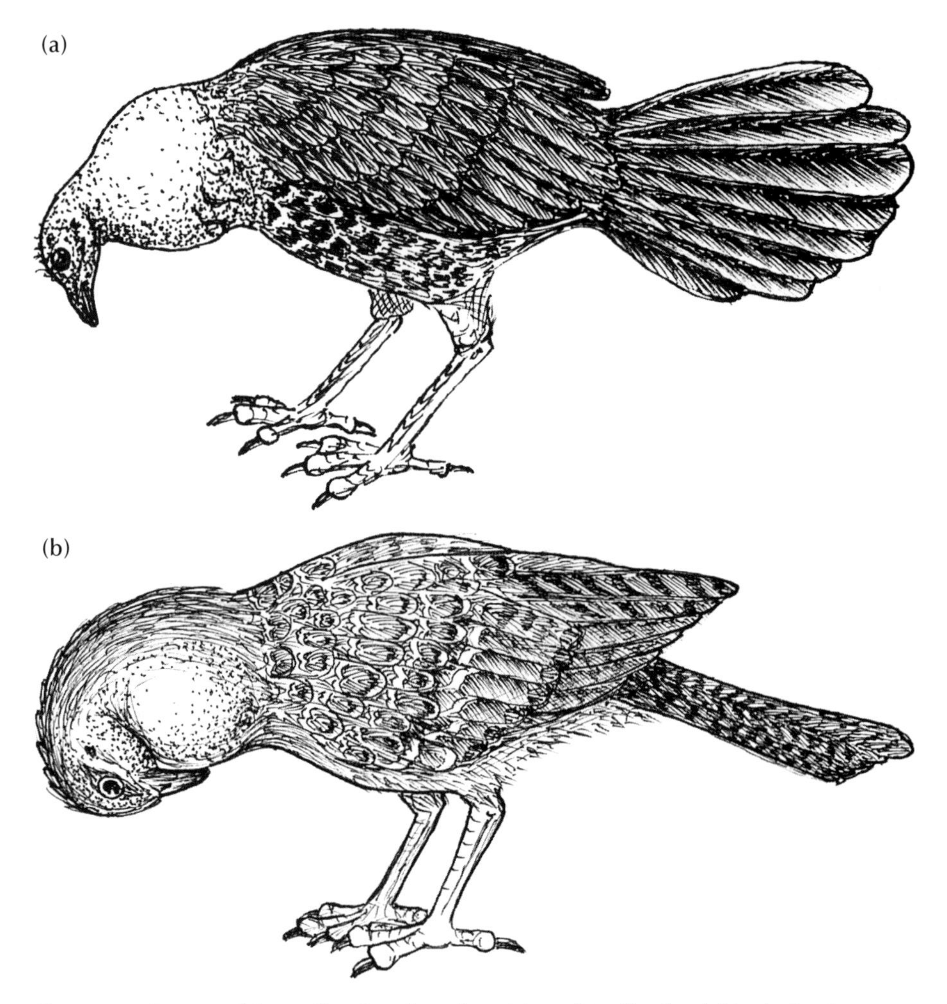


**Figure 3.2 Stance of Australian Brush-turkey (a), and Malleefowl (b), during boom vocalisation.** Drawing: Gwenda White

males fill this hollow sac with air until it is tightly inflated (Figure 3.2) and then force the air out through the nostrils, producing a loud, low-frequency, booming noise of either one ('oom') or more syllables ('oo-oom'). The quality and frequency of these booms varies enormously among males, and both age and size of the wattle appear to influence the sound produced.

Brush-turkeys boom in two main contexts: while males are near their incubation mounds, presumably to advertise their presence to both females and other males; and in competitive situations where rivals are being confronted. Booming is especially common when numerous males may be contesting a valuable feeding location. Similar boom type vocalisations are also known from the Malleefowl which appears to have expandable internal pouches in the neck region.

*Malleefowl*

***Appearance*** While both the Brush-turkey and Scrubfowl are often hard to see because their drab, dark plumage is difficult to detect in the dull light of the rainforest floor, Malleefowl are even harder to detect, even though their habitat is more open and usually well lit. The plumage of Malleefowl blends superbly with the mottled light and yellows, greys and browns of the dense low woodlands in which they live.

In general, the primary underlying colours of the plumage is clear grey above, and light buff or pale orange under the belly and throat. Overlying this simple background is a regular pattern of cinnamon or chestnut 'arrowheads' edged in thick black along the base and white along the top. In 1840, John Gould thought that the larger of these spots resembled eyes (or 'ocellae'), hence his species name of *ocellata*. These patterns are repeated on each wing feather and in smaller motifs on the back, typically arranged in a series of irregular rows that run diagonally up the wings. The pattern on the tail feathers is simplified and resembles jagged zigzags of black on the grey. Running down the throat is a narrow triangle of black 'brushstrokes' ending just below the chest. The top of the head is grey but covered in short spiny feathers, which can be raised at will into a short crest.

There is considerable individual and seasonal variation in the appearance of the plumage patterns. As well as natural differences, the eventual bleaching of the plumage by the harsh sun and abrasion through wear leads to visible differences between birds. This variation has often been used as the basis for distinguishing individuals, although it has yet to be verified for more than a single season.

The sexes of Malleefowl are virtually indistinguishable and almost all measurements that have been recorded were of unsexed adults (Table 3.2). Males tend to be heavier than females and both sexes reach their maximum weight in spring.

***Vocalisations*** In their vocal repertoire, Malleefowl combine the types of calls found in the Brush-turkey (grunts and booms) and in the Scrubfowl (duets), and include some additional calls during egg laying and when alarmed.

Grunts are the pair's typical contact calls. These quiet but often repeated calls are usually produced when the male and female are working together

**Table 3.2 Tarsus length and body mass of adult male and female Malleefowl from three different states**

| | Tarsus | | | | Body mass | | | |
|---|---|---|---|---|---|---|---|---|
| | Mean | S.D. | Range | n | Mean | S.D. | Range | n |
| Males and females | 75.9 | 2.15 | 72.5–80.3 | 13 | 1768 | – | 1520–2-50 | 13 |
| SA males | | | | | | | 2000–2500 | 2 |
| SA females | | | | | 1768 | | 1520–2050 | 5 |
| NSW females | | | | | 1830 | | 1788–1901 | 4 |
| VIC males | | | | | 1947 | | 1650–2270 | 7 |
| VIC females | | | | | 1857 | | 1540–2250 | 6 |

S.D. = Standard deviation.
Source of data: Unspecified sample (males and females) reproduced from Jones *et al.* (1995) with permission; SA sample from Booth (1987); NSW sample from Frith (1959); VIC sample from Benshemesh (1992).

near the mound. When laying eggs, females also utter a 'crooning' call, sounding like 'ooma' or 'whoo how', which they mainly do during their preparation for laying and which become increasingly loud immediately prior to oviposition. The male sometimes utters the same call during the final stages of egg laying. When disturbed, Malleefowl sometimes produce an alarm call that sounds either like a soft 'ut ut' or a loud sharp grunt, but quite often, Malleefowl are silent when disturbed.[4]

Like the Brush-turkey, the most characteristic vocalisation is the boom, which is produced primarily by males near the mound and is regarded as being territorial in function.[4] The production of this sound appears to be similar to that of the Brush-turkey, involving the expulsion of air under pressure. The voice produced has been described as a loud 'uh-uh-uh-oom-oom-oom', which is normally repeated three to five times. When 'booming', the male adopts a particular stance, with the head bent downwards against the heavily inflated breast and the bill pointing backwards (compare the different stances of the Malleefowl and Brush-turkey in Figure 3.2). The call is unexpectedly loud for an otherwise quiet species, being audible for up to 800 m, a feature suggesting that the bird is advertising its presence to other Malleefowl in the vicinity.

The booming call is also the male's part in the pair's duet.[5] As in the Scrubfowl, duetting occurs whenever the male's calls overlap in time with the female's calls. The latter calls of the Malleefowl female can be described as long drawn-out and rising 'waugh, waugh' calls, higher in pitch than the male's boom call. Both male and female can start the duet, but one study found that in 80% of all observations, the female was the initiating partner.[5] Duets are often characteristic of long-term pair-bonds in monogamous birds, and are likely to have a pair-bonding function in the Malleefowl (see Chapter 7 for further discussion of social organization). At the same time, they may also have a territorial function, but this warrants further study. They are not quite as loud as those of the Scrubfowl but can be heard for about 100 m.[5] This may indicate that they also serve to announce a territory claim to neighbouring pairs.

### *Orange-footed Scrubfowl*

***Appearance*** Orange-footed Scrubfowl do indeed have vivid orange legs and feet – which are quite conspicuous when the birds are seen in bright light – but these are really the only colourful aspects of their appearance. The plumage of this smallest of the Australian mound-builders is mainly dark olive-brown in colour with slightly greyer underparts. The face and cheek area is sparsely feathered with grey-brown to yellowish bristle-like feathers through which the skin can be seen. The feathers of the head rise into a prominent pointed crest, although in the deep shade of the rainforest, this is not often seen. Unlike Brush-turkeys, Scrubfowl have no wattle and no obvious tail. Their horizontal tail feathers are so short that the tail hardly protrudes beyond the wings, giving the birds a rounded appearance.

While the Brush-turkey and Malleefowl are both approximately domestic turkey in size, the Scrubfowl is only bantam-like in stature. In the dark environment of the forest floor where these birds spend almost all of their time, they are easily missed or overlooked, especially in areas were they are not used to people. Male and female Scrubfowl are extremely similar in appearance and measurements (Table 3.3). In fact, all available measurements do not distinguish between males and females, making it impossible to establish if any differences in the size of physical features exist between the sexes.

***Vocalisations*** Scrubfowl are much more vocal than the other two mound-builders. Whereas Brush-turkeys have been described as 'occasional grunters' and Malleefowl as 'largely silent apart from low level non-musical notes', Scrubfowl are decidedly operatic in comparison. The vocalisations of Scrubfowl are remarkably loud, repetitive and clearly capable of broadcasting over considerable distances. All the other members of their genus (*Megapodius*) are similarly vocal, and their far-carrying calls are characteristic sounds of the dusk in the many tropical locations where these species occur.

**Table 3.3 Tarsus length and body mass of adult (male and female combined) Orange-footed Scrubfowl from eastern Queensland (specimens from Cairns to Mackay).**

| | Tarsus (mm) | | | Body mass (g) | | | n |
|---|---|---|---|---|---|---|---|
| | Mean | S.D. | Range | Mean | S.D. | Range | |
| Males & females | 73.7 | 3.14 | 69–77 | 1091 | – | 900–1200 | 5 |

S.D. = Standard deviation.
Source of data: Jones *et al.* (1995); used with permission.

These calls may be given at anytime of the day but are particularly common as the birds gather prior to roosting, and often also during the night. The birds seem to be especially vocal during the breeding season when their distinctive 'kee-ow, kee-ow' or 'eukeu-keu-keu-keuw' can be heard frequently. They also produce a series of quieter, closer-range calls (given as 'keurr', 'oowooq-euweuw' and 'peutauw-teurrr'), as well as disturbance calls ('pie-wiet – wiet-iet'), produced when the birds are flushed.

Pairs of the Scrubfowl often call in a duet, that is, in a synchronised manner in which the calls of the male and female overlap. Worldwide, approximately 200 species of birds are known to duet, but most are songbirds; the Australian magpie-lark is a good local example. It is unusual for a galliform to duet, which makes it even more regrettable that we know so little about the function and structure of these vocalisations in the Scrubfowl. Most of what we know today is still based on a study from the 1970s, which described the social behaviour, including calling, of some Scrubfowl pairs in coastal forest just north of Cairns in North Queensland.[6]

When male and female of this species duet, their calls are so synchronised that they sound like the call of one individual. The Cairns study found that the female usually starts the duet with a series of loud clucks, the male then joins in with several 'chirrups', followed by an extended descending double-note described as having a 'desperate mournful quality'.[6] However, it is also possible that the duet is started by the male, as is the case in the closely related Polynesian Megapode, in which pairs produce a duet very similar to that of the Scrubfowl.[7]

Not only is the duet produced in a highly synchronised manner, it is also so loud that it carries far through the dense forests in which these birds live. At night, when the birds call from a tree, their duets can easily be heard up to six kilometres away. Often several pairs duet at once or in response to one another, creating a prolonged braying chorus.

Although understanding the function of the duet will require detailed study, we can assume that it serves similar functions as the duet in the Polynesian Megapode.[7] First, the duet is likely to help paired birds to stay in contact, and to maintain and strengthen the pair-bond. Scrubfowl are almost certainly monogamous, and duets of tropical birds commonly coincide with a long-lasting pair-bond. However, the duets of megapodes are too loud to be directed towards the partner alone and likely to also have a territorial function. Pairs may advertise their location and thus their presence within the territory to neighbouring pairs, thereby avoiding conflicts over territorial boundaries. This suggestion of a territorial function is supported by both the loud volume of the duets and the fact that pairs often respond to each other. Clearly, the duets of the Scrubfowl would be a fascinating and fruitful area for intensive further study.

## Foods and feeding ecology

All mound-builders live primarily on the ground. Whether they are wandering through the damp, steaming coastal forests of the tropics, the cool, rich floor of mountain rainforests, or the dry, harsh mallee, it is among the layers of fallen vegetation and decaying leaf litter that these birds find almost all their food. Except when preoccupied with mound building and maintenance, mound-builders spend most of their time searching through the leaf litter for some of the many food items they are known to consume.

The search for food involves the continual raking and shifting of enormous amounts of leaf litter, humus and other debris such as logs, sticks and even large stones encountered as they progress along the ground. This often results in bare patches on the forest floor, in which the roots of trees may be laid bare. With their remarkably strong legs and large feet and claws, megapodes can toss aside comparatively large stones or decaying logs and contribute considerably to the rearrangement of the forest floor. One Scrubfowl, for example, was observed digging up a stone of almost 7 kg and moving it 70 cm. We have often seen Brush-turkeys determinedly moving large logs and uprooting sizeable rainforest plants. (When they are foraging in suburban gardens, these activities often cause serious damage, resulting in conflict with householders; see Chapter 8).

Most galliformes – such as domestic chickens – and other ground-foraging birds typically rake the ground, then step back in order to search for food in the disturbed ground. In contrast, the long legs of mound-builders allow them to scratch and watch the ground at the same time, and to turn over a considerable volume of leaf litter in a short time. With their strong feet, they can also rip apart dead logs to reach the termites or other invertebrates found inside.

All mound-builders are omnivorous with a very generalised diet. Both adults and chicks feed on a wide range of invertebrates, seeds, fungi, fruit and other parts of plants, depending on what is locally or seasonally available. Among invertebrates, for example, grubs, beetles, worms, the pupae of ants and termites, gastropods, centipedes and millipedes are all eaten. Many fruits are swallowed whole, and mound-builders may therefore contribute to the distribution of the seeds of the plants involved.

In addition, mound-builders also occasionally take larger vertebrates such as skinks and small snakes. Brush-turkeys, the largest of the mound-builders, have occasionally been seen to tear apart frogs and rats with their strong bills, although it is unknown whether they actually hunt such prey or simply discover it as carrion.

The Brush-turkey appears to have the most wide ranging tastes and foraging abilities of the three species. For example, Brush-turkeys have a particular fondness for the succulent tubers and enlarged roots of certain rainforest plants such as Alocasia (or 'Elephant Ears', Family Araceae). Accessing such tubers requires significant amounts of digging and we have observed birds excavating particularly large Alocasia specimens in southern Queensland. Once exposed, the birds consumed the highly succulent and gelatinous interior of the tubers, leaving the rest of the plant untouched. This fondness for succulent underground plants parts has led Brush-turkeys living in suburban areas to seek out similar species in domestic gardens, even when they grow in pot plants. Similarly, Brush-turkeys have been known to consume the underground parts of sugar cane and potatoes.

Brush-turkeys are the only megapode known to occasionally feed on fruit in the forest canopy, sometimes as high as 15 m above the ground. This arboreal foraging ability has allowed the species to feed on ripening bananas in plantations in northern New South Wales. Finally, the Brush-turkey seems to have taken advantage of the foraging opportunities associated with the spread of the noxious weed Prickly Pear *Opuntia* spp. The plant has a large and nutritious fruit and Brush-turkeys learned to appropriate these by flying over the 1–2 m high cacti and grasping the fruit in their feet.

As one would expect, the diet of the Malleefowl tends to be dominated by fruits, buds and seeds of the plants growing within its mallee habitat, although invertebrates are also taken. Nonetheless, there can be substantial differences in the diets of Malleefowl from different locations. For example, Harry Frith characterised the diet of the birds he studied in western New South Wales as primarily granivorous.[8] He determined that over 70% of the food consumed were fruits, buds and especially seeds, the rest consisting of herbs (10%) and invertebrates (17%). In contrast, a female Malleefowl studied intensively by Joe Benshemesh in northern Victoria consumed mainly foliage and flowers with seeds being a relatively minor component.[9] This bird also ate fungi, ants and termites.

Russell Harlen and David Priddel found that a wide range of invertebrates are also eaten by Malleefowl including ants, bees, beetles, cockroaches, dragonflies, grasshoppers, spiders and wasps.[10] In the Little Desert region of Victoria, 'Whimpey' Reichelt has witnessed Malleefowl consuming many different plant species, including plant parts such as flowers, but interestingly, only rarely *Acacia* seed.[11] Surprisingly, lerps – the carbohydrate-rich covering secreted by psyllid insects – have been found to make up about a third of items consumed. These are an excellent source of sugars and were found by Joe Benshemesh to be especially important in the diet of females prior to egg laying.[9] A recent comparison of the foods of Malleefowl from four different locations suggests that Malleefowl exploit an extremely broad range of items leading to highly localised diet, with few species of plants common to different sites.[11]

We know much less about the diet of the Scrubfowl. What is known is limited to casual observations of foraging birds and some stomach contents. This information suggests that animal material dominates, including snails, centipedes, millipedes, beetles, earthworms, ant pupae, larvae and small snakes. The only plant material mentioned is fallen fruits and the seeds of the Tamarind.

Brush-turkeys and Scrubfowl usually live in moist environments or near creeks where they can find drinking water. They also take water that has accumulated in seedpods or other temporary locations. Malleefowl, on the other hand, do not need surface water to survive.[8] Like many Australian species in arid areas, they are adapted to living without water for prolonged periods. They do drink when provided with water in captivity or after heavy rains, but during prolonged dry periods in the wild, they appear to gain enough moisture from morning dew and their food.

## Movements and general ecology

All mound-builders spend most of their time on the ground and usually only move into the trees to roost in the evening. They always roost at night but may also do so for varying times during the day when they are resting. To ascend into a tree, they hop from branch to branch, using their wings to flutter short distances. Individual Brush-turkeys often return to the same roost tree over extended periods, some trees are apparently used for decades. Sometimes, this tree may be some distance from where they spend the day, and their daily schedule is thus strongly determined by moving to the roosting tree in the evening and leaving it to get back to the feeding or incubation mounds in the morning. However, during the breeding season, most males of both Brush-turkeys and Malleefowl appear to roost alone in a tree within sight of their mounds, while the females return to roost in their traditional tree.

Adult Scrubfowl fly more readily when disturbed than the much larger Malleefowl and Brush-turkeys. The latter species fly laboriously and, unless surprised by a human or predator at close quarters, usually run away rather than take to the air. If disturbed while roosting, Brush-turkeys seem to drop, rather than fly, from their perch, gliding away only when near the ground.

Chicks of all three species are far more capable of covering larger distances in the air than the adults. Hence it is usually the young birds that colonise new areas, sometimes crossing considerable distances over unsuitable habitat or water.[12] This ability supports the suggestion that such dispersal by young chicks originally led to the establishment of the many new populations of *Megapodius* on islands throughout the South Pacific.

On the ground, mound-builders can run rapidly, supported by their strong legs and, in the case of Brush-turkeys, are apparently aided by their vertically aligned tail which acts as a rudder. When pursued by a predator, they tend to run rather than fly, which can be detrimental when trying to evade a fox or cat, or even a car. They also tend to run along obstacles such as fences, rather than flying over them. In the case of Malleefowl, this means that the birds run along the fences erected by rabbit hunters, resulting in many being caught in the rabbit traps.

## Moulting

Like all birds, mound-builders must replace their feathers on an annual basis. However, unlike most birds, which shed the majority of their feathers fairly rapidly over a period of days or weeks' time, mound-builders do so gradually over several months. This type of gradual moult (the so-called 'staffelmauser') means that they shed only 1–3 feathers in the wings and maybe 1–2 in the tail at any one time. They then wait until the replacement feathers have grown to much of their final length before the next set of old feathers are shed. The timing of moulting in most of the species is still poorly understood, although it appears that this occurs primarily in the non-breeding season, with only the occasional feather being replaced during the breeding period.

## Life-span

It is not yet known how long mound-builders live in the wild, and there are few reliable age records from captive individuals. Our observations of wild Brush-turkeys indicate that adults live for at least nine years, as two males and one female banded by us were seen continuously in the same patch of rainforest eight years later; these were at least one year of age when banded. Both of these males were engaged in constructing mounds for at least those eight breeding seasons. A similar observation was from one male recognisable by a bent beak, which inhabited the same suburban property for at least seven years before he disappeared. Several pairs of Malleefowl have been observed by 'Whimpey' Reichelt in the Little Desert regularly and for several decades. Two known individuals, a male and female, are now at least 30 years of age and are both still actively breeding.

# 4
# THE MOUND

*The scrub-fowl is a mighty builder, using 'cartloads' of dark, sandy soil mixed with decaying vegetation, twigs, sticks, etc. Sand and shells are the principal building material when the egg-mound is near the seashore. (C. Barrett 1931: 117)*[1]

*We think of invention as the sole prerogative of man. Yet nature was perfecting her inventions long before man was born. For instance, there are the incubating nest-mounds of the Australian megapodes, or mound-builders, predating man's invention of the incubator by an incalculable period of time. (A. Russell 1944: 83)*[2]

All of the Australian megapodes are mound-builders, in contrast to the 'burrow nesters' in some areas outside Australia (see Chapter 1). Megapode mounds are large heaps of decomposing organic material, which contain whatever the birds scratch together at the mound location: leaf litter, fibrous roots, smaller pebbles and stones, branches, some top soil, etc. While the most important material will be that which can be broken down by micro-organisms, a lot of other objects may be inadvertently gathered up in the process, even though this latter material may play no part in the incubation function of the mound. Suburban Brush-turkey mounds, for example, often contain artificial artefacts, including soccer balls, drink cans and silver spoons. In developed areas, Brush-turkeys frequently take over compost heaps or piles of lawn clipping and convert these into an incubation mound.

Scrubfowls and Brush-turkeys tend to construct their mounds in moist gullies or woodlands, and therefore have plenty of suitable material available for mound construction, such as fallen and decaying vegetation, soil, sticks and twigs. Malleefowl, on the other hand, inhabit arid areas where vegetation cover is considerably less dense, often with a thin layer of leaf litter. The major portion of their mound thus consists of the light sandy soil of the mallee country, with only limited amounts of organic matter concentrated into the central core of the mound.

The lack of organic building material may also be the reason why Malleefowl construct the smallest mounds of all three mound-builders. On average, these are 60–90 cm high and 3.7 m wide. One estimate of a Malleefowl mound's mass was 3400 kg.[3] The Brush-turkey mounds are, on average, 0.85 m high (± 0.15 m S.D.) and 3.6 m wide (± 0.6 m S.D, n = 25), although some may reach 1.3 m in height. In eastern Australia, Brush-turkey mounds consist of an average of 2268 kg of moist soil and litter per mound (n = 5, max 4000 kg)[4] but on Kangaroo Island, mounds contained up to 6800 kg of material.[5] By far the most impressive mounds, however, are those of the Scrubfowl. In northern Queensland and the Northern Territory, these mounds can reach heights of 5 m and a diameter of 13 m, although most mounds are of a more modest height, around the 1–2 m mark.[6]

When considering the size of megapode mounds, one has to keep in mind that there is considerable variation in mound size within each species, for two main reasons. First, mounds change in height as the birds either flatten the top or heap up more material during their daily maintenance work, depending on the prevailing weather conditions at that time. This makes it difficult to describe the 'average' height or 'typical' shape of a megapode mound. Second, in all three species, some birds construct their mounds from scratch, whereas others use the same mound location in successive years and thus build their new mound on top of remnants of an earlier one. The original mound will have been reduced to a low pile of compressed soil and will no longer produce heat, but will add to the overall height of the new construction.

Scrubfowl in the Northern Territory, in particular, are known to reuse the same mound season after season, resulting in enormous structures.

For example, one mound was used for over 30 years, while another particularly impressive mound was used for more than 50 years. The large mounds of Scrubfowl have shaped the landscape in Northern Australia to such an extent that they have become 'fossilised' and remain permanent parts of the environment. One study has dated such fossilised Scrubfowl mounds at more than 1500 years old![7]

## Mound construction and maintenance

After selecting a site, the development of a functioning incubation mound requires two phases of activity: the initial gathering of material at the site ('construction'); and the longer-term process of assessing temperatures, adding or removing new material, tending and generally maintaining the structure ('maintenance'). Mound construction and maintenance serves two main purposes: the utilisation of the heat produced through the decomposition of organic material by micro-organisms, and the enhancement and prolongation of this decomposition by controlling the temperature and moisture levels within the mound. Mound construction in the three species differs in terms of mound composition, the size and repeated use of mounds, and the extent to which males and females are involved in this mound work.

In Brush-turkeys, only the males build and tend the mound. When raking together material for the mound, they work backwards, vigorously tossing the material far behind them as they proceed. At first glance, their actions resemble those of a leaf blowing machine. First, they rake together the loose material gathered together on the forest floor into a narrow strip that can extend up to 100 m or more from the mound. At the right times of the year (often late winter or early spring), a careful observer can find the location of a well hidden mound by following the trails of leaf litter on the ground. The males may work on these trails for up to two weeks before starting to rake the material onto the mound, and they shift on average about 56 kg of material each day.[4] During this phase, males spend most of their day raking and are not easily distracted from mound building. In suburban areas, they often move their material across roads and are undeterred by fences, building a ramp of material against one side of the fence and kicking the material over the hurdle and onto the other side.

Vigorous raking and gathering of material into a pile is typical of the first weeks of the 'construction phase'. Once the mound is established, the 'maintenance phase' begins and the male now also works on the mound itself. He starts to mix the material in the mound in order to generate sufficient heat for incubation. This involves raking surface material from one location on the mound to another, digging deep holes for vertical mixing and testing the temperature at regular intervals.

At the same time, he continues to rake fresh material from the surrounding substrate onto the mound, often completely denuding the surrounding area of any leaf litter. By this time, he has raked an average of 126 $m^2$ and the ground within a radius of up to 150 m from the mound has been raked bare.[4] This material, now concentrated into a single pile, includes all of the fallen seeds and fruits as well as the humus layer of the forest floor, which is the primary source of accessible nutrients needed by the local plants. Hence, the mound construction activities of Brush-turkeys and other mound-builders may have a profound influence on the structure of the forest itself. Indeed, these features strongly suggest that mound-builders may be regarded as 'ecosystem engineers' or 'niche constructors', species which are able to significantly alter their environments to their advantage.

All of the activities associated with the maintenance of the mound as a heat source are associated with facilitating the growth of healthy and expanding populations of micro-organisms and heat-adapted fungi. As they metabolise plant cells, these myriads of minute organisms produce the heat that is harnessed by the mound-builders; the 'engineering' of these birds is thereby occurring at both a macro and micro level.

Approximately one month after the onset of mound building, Brush-turkey females first start visiting the mounds for egg laying, although this period may be delayed if insufficient rain has fallen to facilitate the decomposition of the organic material. The females' activity on the mound is limited to digging tunnels down into the mound to assess the internal temperatures, egg laying and depositing the eggs; otherwise, they do not participate in any 'mound work'. The males usually close the tunnels opened by the females as soon as these have left, to prevent excessive loss of heat.

Male Brush-turkeys spend considerable amounts of their time rearranging material in response to their assessments of the internal temperature of the mound (see pages 39–42). For example, if they perceive the mound to be too hot, they open up the surface, producing a flat or slightly saucer shaped top. Alternatively, they rake more leaf litter onto the top of the mound when the temperature is too low or when rain approaches.

Scrubfowl differ from Brush-turkeys in that both pair partners are involved in mound building and seem to share the workload almost evenly.[8] In one intensively observed pair, the male spent twice as much time as the female collecting new building material, but both took equal shares mixing the material on the mound on a daily basis and digging holes for testing the temperatures.[8] Some Scrubfowl mounds are also used by several pairs simultaneously, a behaviour not known from the two other Australian mound-builders.[8]

In Malleefowl, both pair partners are involved in mound building and it is less clear who is involved in the majority of the work. Some observers state that the male does most of the work, with only occasional assistance from the female, while other studies found that the female also contributes significantly to this work, even though the male spends more time on, or close to, the mound and is the first to visit it in the morning. In a recent, long-term study, Jessica van der Waag found that females are usually more wary and more likely to leave the mound when people are present, which may explain the observation that females only occasionally assist with the work.

In their harsh, arid environment, Malleefowl need to expend considerable energy in providing suitable incubation temperatures, especially during times of drought, floods and soaring heat. When starting a new mound, they select a sandy rise and excavate a depression in the ground. They then start scraping whatever organic matter they can find into piles near the depression, and this material is then left lying until saturated by rain. Once wet, the leaves are transferred into the hollow depression, usually some time between June and August, where it is left exposed to the weather. Only once the winter rain has saturated the leaves do the birds construct the mound proper by heaping sand on top of it. The dry sandy layer acts as effective insulation, trapping the heat and moisture within the mound in the inner 'egg chamber'.

Once constructed, a Malleefowl mound needs considerable daily labour to ensure stable and suitable temperatures, primarily because any access to the egg chamber necessitates the moving of the sand layer. Early in the breeding season, the decomposition of the leaf litter produces sufficient heat for incubation, but the sand layer must be continuously replaced to avoid the loss of heat and moisture from the egg chamber. Such moisture is essential for decomposition. At regular intervals, the birds check the temperature inside the egg chamber, mix the organic material or add more leaf litter. The temperature balance is adjusted by changing the amount of sand heaped on to the mound. If the egg chamber is too hot, sand is removed to allow heat to escape, while sand is added if it is too cool. In order to reach the egg chamber for temperature testing or egg laying, up to about 850 kg of sand must be removed and replaced each time.[9]

The work of the Malleefowl becomes even more sophisticated during times when the decomposition of the leaf litter does not produce enough heat for incubation, such as the last months of the breeding season. This is typically mid to late summer when most of the organic material has been exhausted. The birds then begin to utilise the sun's heat for incubation, but they need to do so carefully, as extensive exposure to the often extreme heat of this time could easily endanger the eggs.[10] On the morning of a sunny day, they scratch away most of the sand that covers the egg chamber, leaving only a layer 2–3 cm above the eggs. This enables sufficient solar radiation to penetrate to the level of the eggs. As the day progresses, the birds gradually refill the depression, replacing the soil layer after layer, while constantly testing the temperature with their beak. By the time the sun is setting, all of the sand will have been heaped back onto the mound. Day after day, this rather laborious yet intricate procedure is repeated, though only on sunny days.[5] This remarkable behaviour has evolved in Malleefowl in response to the extreme aridity and unpredictable rainfall patterns in their particular surrounding. No other mound-builder spends so much time and energy on ensuring that their mounds are functioning properly, primarily because none of the other species lives in such a seemingly unsuitable habitat.

In summary, all three mound-builders regulate the heat production inside their mounds by removing or adding material throughout the breeding season. How they monitor mound temperatures is not yet fully understood. Most likely, both males and females possess a temperature sensor in the palate or tongue, as all species have been seen to regularly take a beak-full of substrate while working on the mound. The exact site and structure of such a possible sensor, however, has yet to be determined.

Clearly, the construction and subsequent maintenance of an incubation mound is a very big job. It is, therefore, no surprise that sometimes a Brush-turkey mound is taken over by a male that did not build it, forcing the constructor to start again elsewhere or give up. Such usurpation enables that male to avoid the huge time and energy expenditure associated with mound building. However, most males (in Brush-turkeys) or pairs (in Malleefowl) manage to defend their mound throughout the breeding season. They are highly territorial and vigorously expel all other males or pairs seen close to their mound.

If members of the same species are known to use the same mound, what about different species of megapodes? In the jungles of Papua New Guinea, indigenous people have long described a type of mound 'sharing' where the eggs of more than one species are found within the same mound. In the only case investigated by a scientist (P. Dwyer), a species of *Talegallus* and a scrubfowl (*Megapodius*) laid their eggs in the same mounds in Papua New Guinea. Among the Australian species, this phenomenon has so far been described by indigenous people: an aboriginal elder (W. Canendo) from the Atherton Tablelands of North Queensland reported that he occasionally found eggs of both Scrubfowl and Brush-turkey within the same mound. These observations raise some intriguing questions about how common this strategy is, which species build the mounds and what the advantages and disadvantages for each species are.

## Mound temperatures

All three mound-builders incubate their eggs by utilising the heat produced through the decomposition of the moist organic material incorporated into their mounds. Malleefowl are a notable exception in that they also utilise solar heat for incubation. The previous description shows how diligently they are able to manipulate heat production in their mounds by removing or adding building material.

A large army of bacteria, fungi and small invertebrates living inside the mounds is responsible for the decomposition of the organic material. These organisms increase their populations and activity whenever the mound is moist enough or when fresh organic material is supplied to them. Their increased activity leads to faster decomposition, and thus to the production of more heat for incubation. Hence, to maintain suitable conditions for these micro-organisms, megapodes invest considerable amounts of work in their mounds, as described previously.

All mounds start as a growing pile of freshly gathered, damp leaf litter and soil, along with whatever micro-organisms are living naturally in this material. Obviously it takes some time before the populations of these animals and fungi begin to grow, and this will be affected by the local temperatures, moisture levels and the type of material incorporated into the mound. Throughout this time, both males and females will be checking the temperatures deep within the mound, waiting for the time when these are both stable and suitable for incubation. The internal temperature of one Brush-turkey mound that was carefully monitored began to rise almost immediately, peaked at about 39.8°C after about 28 days before stabilising at about 33°C after 6 weeks[11] (Figure 4.1).

What are the incubation temperatures in megapode mounds? In general, megapode eggs tolerate a wider range of incubation temperatures than

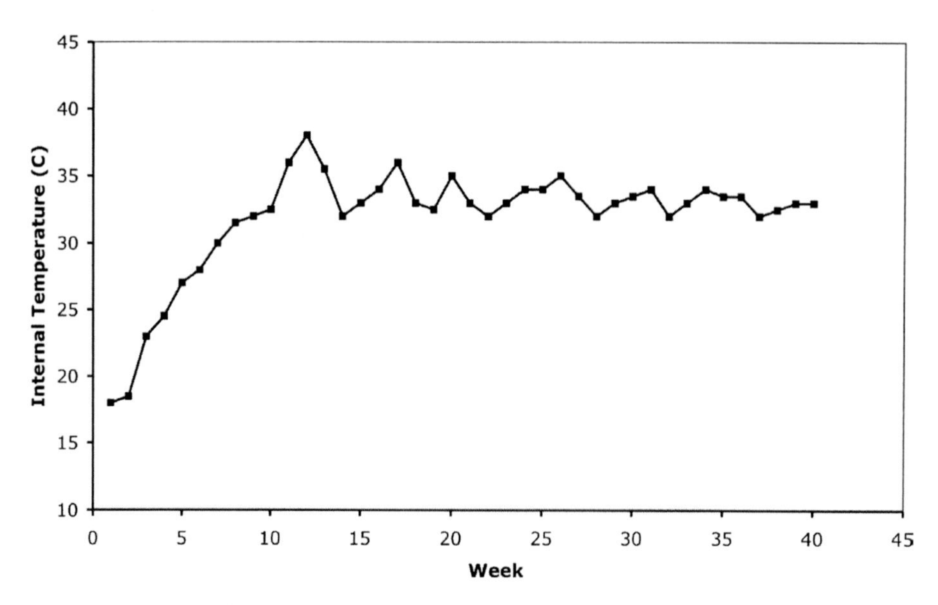


**Figure 4.1 Internal mound temperatures recorded weekly for 40 weeks at Mt Tamborine, south-east Queensland (Week 1 = May 12).** Data from Darryl Jones.

other birds' eggs. In Malleefowl mounds, for example, incubation temperatures vary between 27° and 38°C, but most eggs are incubated at 34°C.[10] Temperatures of 35–39°C have been recorded for a few eggs in Scrubfowl mounds, but so little data exists on this species that future studies are likely to detect a wider range of temperatures. Ann Göth's studies of Brush-turkey mounds have revealed that temperatures next to eggs ranged anywhere from 27° to 37.3°C,[12] yet the mean temperatures from three separate studies were remarkably similar: 33.2°C,[12] 33.3°C[4] or 33.0°C.[13] Figure 4.2 shows the distribution of incubation temperatures for 339 eggs obtained over two breeding seasons, from 16 different mounds on the Central Coast near Sydney.

This information shows that there is a considerable variation in incubation temperatures found between species, and within mounds of the same species. What could cause this variation? Many factors may be involved, among them the time of year, the prevailing climate of the local area, the mound location, the size of mounds and their composition, and the males' ability to build mounds.

The period over which mounds are constructed and maintained usually lasts for many months, and hence, the time of the year when temperatures are recorded can affect the results of such measurements. Figure 4.1 shows

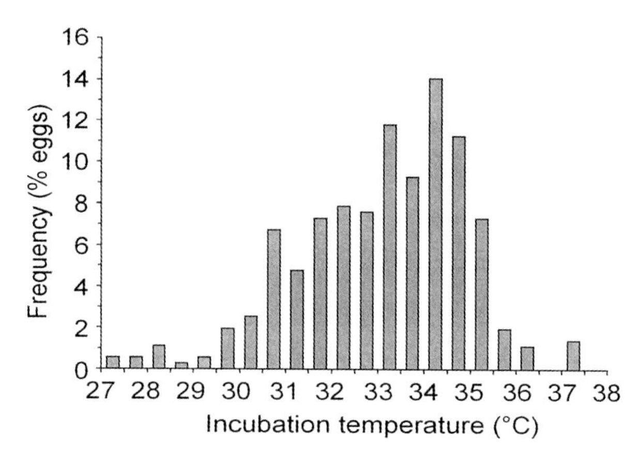


**Figure 4.2 Incubation temperatures of 339 eggs obtained from 16 mounds over two breeding seasons.** Reproduced from Göth (2007)[12] with permission from Blackwell Publishing.

that when mounds are first built, temperatures can rise to very high levels, close to 40°C, before stabilising at about 33–34°C, a suitable level for incubation. Furthermore, mounds that are built early in the season – which is typically in winter – heat up relatively slowly, while those built later in warmer times heat more rapidly.

Several studies of Brush-turkey and Malleefowl have also demonstrated that the temperatures next to eggs often differ from those measured elsewhere in the mound. It is thus important to measure the temperatures in those regions where the eggs are found, not elsewhere. However, even if temperatures are only measured next to eggs, in different Brush-turkey mounds and at the same time of year, they sometimes differ considerably between mounds. A comparison of 16 different mounds made in the same month, all located in the same area, revealed that in one mound, the range of temperatures next to eggs was 9.1°C (n = 29 eggs), while in another, it was only 1°C (n = 5 eggs).[12] Seasonal effects, as described above, can hence not fully explain the observed temperature variations, at least in Brush-turkey mounds.

Mound locations (described in more detail below) also affect incubation temperatures. In wet gullies, at the edge of creeks or lakes, or under shaded trees, mounds are likely to be moister than in more exposed or drier locations, and this is likely to affect temperature production. High water content inside mounds increases the activity of the micro-organisms and thus the production of heat. At the same time, however, wetter mounds lose heat more quickly, as high water contents are linked to increased thermal conductivity.[9,14] In addition, wet mounds are less permeable for various gases that affect embryo development, in particular carbon dioxide and oxygen.[14] High levels of carbon dioxide and low levels of oxygen can be fatal to the embryo, and embryos inside mounds built in more humid climates face a higher risk from exposure to lethal concentrations of these gases. Mound-builders, therefore, face the challenge of keeping moisture levels inside their mounds at a level that favours heat production but without endangering the developing embryos. How they do so is not yet known, but they may adjust moisture levels by changing the shape, composition and size of the mound.

Mound size does not necessarily predict mound temperature, and the old saying 'the bigger the better' does not always apply to megapode mounds. We have sometimes found suitable temperatures around 34°C and a high number of eggs in Brush-turkey mounds that were only approximately 80 cm high and 2 m in diameter. On Kangaroo Island, small Brush-turkey mounds (1 m high, 3.6 m diameter) also contained several eggs.[5] To investigate these factors in detail, Roger Seymour constructed artificial Brush-turkey mounds to establish their minimum size as suitable incubators. He found that these needed to have a minimum height of 75 cm and minimum diameter of 2 m to produce sufficient heat.[5] This minimum size is much smaller than many Brush-turkey mounds in the wild.

Heat production and mound size are also linked to the materials used in the composition of the mound, in particular the type of leaf litter incorporated into the mound-builder's nest. As many suburban gardeners know, *Eucalyptus* leaves decompose at a slower rate than, for example, leaves of most rainforest trees, and they also produce less heat. Mounds built in areas dominated by *Eucalyptus* trees therefore need to be of a larger size to produce sufficient heat.[11] Similarly, mounds may also have to be of a larger size in areas where the birds are forced to include large amounts of branches, twigs, small logs, gravel or other material in their mound, which do not decompose much at all.

Mound size also determines how exposed the eggs are to climatic events, such as sudden changes in outside temperatures, storms and strong rainfalls. A larger mound provides a larger buffer between the eggs and such unfavourable outside conditions. In some Brush-turkey mounds, for example, temperatures dropped to 8–12°C when thoroughly saturated after rain.[4] On the other hand, larger mounds require more digging effort from the birds, to mix the material, maintain suitable temperatures throughout the mound and to dig tunnels for egg laying. In summary, the size of the mound is likely to have numerous influences on incubation temperatures, but further studies are needed to determine their importance. Such studies need to consider the difficulties with measuring mound size, as described on page 34.

The discussion so far indicates that numerous factors can affect the incubation temperatures within a mound. Hence, it seems likely that individual experience is of great advantage when the birds need to decide where to site their mound, how large to build it, when to open or close it and when to add or remove material. It may take young males some years before they are capable of constructing a functioning incubator; the descriptions of 'practice' mounds mentioned by some of the early naturalists may represent the efforts of inexperienced birds. In southern Queensland, Darryl Jones followed six sub-adult Brush-turkeys in the wild and found that although these males did build apparently suitable mounds, none received eggs in their mound during their first year, and only one was successful in his second year.[15] Brush-turkey males often attempt to build mounds at an early age – in captivity as early as the age of 4.5 months,[16] but many of these are abandoned soon after construction. This could be due to the lack of sufficient mound-building experience, or due to other, more dominant males expelling these younger rivals from the area.[15]

## Mound site locations

Megapode mounds, in general, are located in a wide range of habitats, although the typical location is in a wooded environment with a dense canopy cover. Such locations favour mound building because they provide a good supply of leaf litter and enough shade to avoid the loss of moisture from the mound through desiccation. In Brush-turkeys, males usually avoid areas with lots of *Eucalyptus* trees, most likely because their leaves do not decompose quickly enough to provide incubation heat.[4] On the other hand, Malleefowl living in arid areas typically dominated by hard and resistant leaves such as *Acacia* and *Eucalyptus*, are unable to avoid these leaves and incorporate them into their mounds. Again, unlike 'typical' megapodes, Malleefowl tend to opt for more exposed mound sites, as these will provide greater exposure to the direct sunlight that will be employed later in the season.

Factors such as the amount of shading or sun exposure, vicinity of the mound to water or the slope of the ground all affect internal mound temperatures. No study has yet looked in detail at how these factors are correlated with the number of eggs found in a mound, but our observations indicate that they are of importance. During the first half of a very dry breeding season, for example, only two mounds in a large Brush-turkey population near Gosford produced eggs. One was located at the bottom of a wet gully, where the ground was still moist, the other at the edge of a lake, where small waves occasionally touched the base of the mound. Only these mounds contained eggs and seemed to be moist enough to decompose at a sufficient rate. In the second half of that particular breeding season, however, torrential rain flooded both of these mounds with the result that all remaining eggs died. During this time, other mounds located elsewhere did produce eggs. This example shows that, depending on weather condition and season, different locations can confer an advantage.

Where Brush-turkeys or Scrubfowl build their mounds on the floor of rainforests, they appear to carefully consider environmental features that relate to the supply of suitable mound material. For example, Brush-turkey and Scrubfowl mounds are often close to the base of large trees, especially Moreton Bay Fig trees *Ficus macrophylla* and Tamarind trees *Tamarindus indica* for the species respectively.[5] In Brush-turkeys, this has been explained as a means of ensuring both a shady spot and a good supply of future leaf litter.[17] Both species also often build their mounds around the base of a large dead tree stump, possibly because such stumps give stability to the mound or, if rotting, produce additional heat.

Brush-turkeys, however, can also be remarkably flexible in mound locations when necessary.[3] In suburban environments, for example, we have found functioning mounds situated in sites fully exposed to the sun, and constructed largely of lawn clippings and garden plants.

Scrubfowl also choose a wide range of different types of mound locations. One account of their typical nesting location describes this as 'just above the high water mark on the beach', whereas a second states that mounds are usually located among dense vegetation, but sometimes at the edge of the jungle.[9] Overall, most Scrubfowl mounds seem to be built in dense forest where the birds find abundant leaf litter for construction. One early account from Dunk Island in Queensland even states that these birds lay their eggs in burrows in sandy beaches, where the solar energy incubates the eggs. Such practice is common among megapodes outside Australia, but not known from any other location within Australia. It is also likely that in this species, different types of mounds have different advantages, depending on the prevailing climate during the breeding season.

As already indicated, Malleefowl appear to be far more constrained in their ability to choose locations for their mounds, as they are confined to what may appear, to the human observer, as endless flat and monotonous mallee. Harry Frith observed that the drainage capacity of the soil was a key determinant for the choice of mound site, with well draining sandy soils preferred over loam and clay. In other areas, however, these features appeared not to be important.[10] A more comprehensive study of Malleefowl distribution in Western Australia, conducted recently by Blair Parsons, found that these typically occur on well drained soils in either mallee or *Acacia/Allocasuarina* shrublands, and often on gravel-type soils. Mounds were most often found in fairly dense vegetation with a low canopy (<6 m) and in areas where litter was abundant and ground cover, such as grasses, was sparse or even absent.

One question as yet unresolved is whether megapodes choose their mound location not only to provide suitable incubation conditions, but also to maximise the survival chances of their chicks. Survival of Brush-turkey chicks is highest in areas with dense ground cover, and indeed, many mounds of this species seem to be built near areas of *Lantana* or other thickets.[11] Similarly, a study of mound site locations in Scrubfowls also showed that 70% of all incubation mounds were located in rainforests with dense vine thickets.[18] It may thus be that parental care in mound-builders goes beyond the physical effort of mound building, to include choosing sites that favour the survival of chicks after hatching.

Whether mounds are built at the same location year after year differs between regions. In southeast Queensland, most (90%, n = 31) Brush-turkey males built entirely new mounds each season, even though they often return to the old earthen base of a mound used some years previously.[4] On Kangaroo Island, however, mounds were used for several years in succession[13] as were mounds in suburban areas north of Sydney. The reasons for this difference are not yet known, but may relate to parasite infestation. When digging up Brush-turkey mounds in southeast Queensland, we were often infested with tiny mites that apparently lived in huge numbers in the mounds. These minute animals burrow into the skin and cause considerable itchiness for several days. It is highly likely that these mites also plague the Brush-turkeys, including the chicks as they dig their way out of the mound. In contrast, excavation of mounds further south near Sydney, never lead to such mite infestations, probably because of the cooler air temperatures (personal observations). By avoiding the re-use of a recently used mound, the sub-tropical Brush-turkeys may attempt to minimise an accumulation of these mites.

## Breeding seasons

Although all three mound-builders spend more time in behaviour associated with reproduction than most other birds, breeding activities occur only during certain months of the year. The onset of egg laying is strongly dependent on rainfall, as only humid leaf litter decomposes at a sufficient rate to produce the heat for incubation. In addition, it is likely that after rainfall, females find more invertebrates, such as earthworms, in the leaf litter and soil. This high protein food is needed for the production of their large eggs. The strong relationship between egg laying and rainfall in megapodes was emphasized by other authors years ago and often caused us to wait impatiently for rain when trying to establish research projects.

In Brush-turkeys, mound construction starts some time between May and August, and is most likely triggered by rainfall. In southern Queensland, mound building seems to follow the first rainfalls in excess of 100 mm occurring after May,[4] with egg laying typically commencing four or more weeks later. Both egg laying and mound maintenance activities gradually wind down some time during the period December to February. Before the start of the next season, mounds are usually abandoned and the birds moult their feathers. However, our observations in Queensland and New South Wales show that occasionally, males may also work their mounds outside the normal breeding season, without any signs of female egg laying. Most likely, these male-only activities are triggered by rain and unusual warm temperatures during that time, providing suitable conditions for mound work, despite a lack of interest by females.

Malleefowl work on their mound almost year-round, even though they only produce eggs during a confined breeding season. A long-term study by David Priddel and Rob Wheeler found that roughly 90% of eggs were laid between the last week of September and the first week of January, with some other eggs being produced from mid August and until mid February.[19] During drought, the birds may not breed at all. Malleefowl prepare the mound for the next season some time between March and June, by removing old material from within it. Like Brush-turkeys, this activity is usually triggered by the autumn rains. During the subsequent months, the birds are occupied with gathering leaf litter from around the mound, moving this material into the mound and heaping up sand. Like the Brush-turkey, mound building activities takes place over a period of several months before egg laying and should not be regarded as the actual 'breeding season'; the period involving the tending of mounds as an incubator of eggs is considerably shorter in all species. While Malleefowl over a large area are generally synchronised in their mound activities, individual Malleefowl pairs may respond at different times to local conditions. For example, it is not unusual to find mounds within a single location to be at different stages of construction.

Little is known about the breeding season of the Scrubfowl, but this seems to differ greatly between different regions. In North Queensland, egg laying seems confined to the period between the first rains and the wet season proper – from September to approximately February.[8] In the Kimberleys, active mounds were found between June and October.[20] One pair studied intensively near Cairns produced eggs over a period of 20 weeks, between August and January.[8] In general, these birds do not usually seem to breed during the wet season, presumably because the intense rains make normal mound activities impossible.

## Mound-builders as weather prophets

Hopefully our discussion so far will have shown how strongly breeding success depends on rainfall. It seems clear that mound-builders will be advantaged by the ability to adjust their activities to rainfall events. For example, Malleefowl close their mounds immediately once the rains have moistened their collected leaf litter, Brush-turkeys activate their mound-building quickly after the first rains of winter, while Scrubfowl quit altogether once the heavy wet season is upon them.

Reacting to rain is one thing; 'predicting' when the rain will occur is quite another. Nonetheless, several (so far anecdotal) observations indicate that all three mound-builders do, indeed, 'sense' the onset of rainfall before it actually starts. Bushmen working in coastal Queensland used to regard Brush-turkeys as rain prophets, claiming that the males start building their mounds at the approach of wet weather. During the breeding season, Brush-turkeys open up the mound in light rain, letting the water penetrate the material to speed up decomposition.[4] Strong rainfall, however, would flood the mound and drown the eggs, or reduce the mound temperature considerably. At the approach of such heavy rainfall, males move more material from the base and side of the mound onto the top, creating a conical shape from which the water runs off more easily. Similarly,

Aborigines on the Atherton Tablelands in Queensland believe that both Brush-turkey and Scrubfowl can predict rain, as they start heaping material on top of their mound several days before the onset of rainfall (Wayne Candendo, personal communication).

Harry Frith's observations of the activities at one particular Malleefowl mound showed that, two days before heavy rain, the female heaped material onto the mound and thatched it with branches from the surrounding area, apparently to lessen damage by rain or wind and to facilitate water run off. At the beginning of a heat wave, she built up the mound exceptionally high, and the thick buffer of sand prevented the heat from entering the mound. As a result, the surface temperature on the mound was 55°C compared to 36°C inside the egg chamber. How these megapodes sense upcoming rain or heat waves, we do not yet know.

While egg laying, a female Australian Brush-turkey (in front) spreads her wings widely. Photo Ann Göth.

An Australian Brush-turkey male resting in a sunny spot on the rainforest floor.
Photo: Ann Göth.

**A pair of Malleefowl on their incubation mound, with the male in the front.**
Photo: Jessica van der Waag.

**A female Malleefowl on the incubation mound.** Photo: Jessica van der Waag.

Malleefowl chick, one day old. Photo: Jessica van der Waag.

A freshly-hatched Malleefowl chick, found when unearthing a mound.
Photo: Jessica van der Waag.

A two-day old Australian Brush-turkey chick emerging from the surface of the incubation mound. Photo: T. Anthony Davis.

A freshly-hatched Orange-footed Scrubfowl chick, found when unearthing an incubation mound and still attached to the inner egg membrane.
Photo: Stan Breeden.

The smallest of the Australian mound-builders, the aptly named Orange-footed Scrubfowl, uses its large feet to forage through the leaf litter on the rainforest floor.
Photo: John Manger.

A four-month-old Malleefowl juvenile with its crest erected.
Photo: Jessica van der Waag.

Two six-month-old Malleefowl juveniles, tracked together in a radio-tracking study.
Photo: Jessica van der Waag.

A six-month-old Malleefowl resting among the Mallee. Photo: Jessica van der Waag.

**A Malleefowl mound largely comprised of gravel and stones.** Photo: Jessica van der Waag.

**A pair of Mallefowl mound tending their mound which is largely comprised of sand.**
Photo: Jessica van der Waag.

An Orange-footed Scrubfowl mound in coastal monsoon forest near Darwin.
Photo: Ann Göth.

A suburban Brush-turkey mound constructed against a fence. Photo: Ann Göth.

Some suburban residents are very patient when it comes to accepting Brush-turkey mounds in their backyards. Photo: Ann Göth.

The plumage of mound-builders are mini-ecosystems for a range of specialised ectoparasites including these feather mites from the wing of an Australian Brush-turkey. Photo: Heather Proctor.

# 5

# ABANDONED EGGS

*We need not concern ourselves with wondering whether these mound-builders have discarded the habit of brooding or simply have failed to acquire it. At any rate, they have not spared themselves labour by adopting this curious procedure, for a great deal of work, highly skilled in nature, is necessary. (A. Chisholm 1934: 44)*[1]

*Expulsion of the egg apparently places a severe strain on the bird, particularly in hot weather, and she may lie at the bottom of the hole for several minutes. Having recovered, the female usually retires and takes no further interest in the egg. (H. Frith 1962: 95)*[2]

Megapode embryos need to cope with a highly unusual array of conditions, many of which would normally be fatal to other birds. The extreme temperature fluctuations or high concentrations of fatal gases in the mound, for example, would certainly kill the embryos of any other birds. Yet megapode embryos withstand such conditions, because of several special adaptations. This chapter describes these adaptations and answers the following questions: (1) What do eggs look like inside and outside?; (2) How many eggs are laid and how many of them actually hatch?; (3) For how long are eggs incubated?; and (4) What is the embryo's response to changes in gas pressure and incubation temperatures? The last question is of particular interest because it has recently been discovered that megapodes are the only birds in which incubation temperatures are known to affect sex ratios.

## Megapode eggs and their special adaptations

Megapode eggs are almost symmetrically oval in shape, sometimes slightly elongated. In Brush-turkeys, they are pure white, whereas in Scrubfowl and Malleefowl eggs are pinkish or brownish in colour, at least at the time they are laid. In the latter species, the pigment is present only in the outer coating of the egg which is easily rubbed off, exposing the white surface beneath. This means that older eggs can be distinguished from the unspoiled pink of freshly laid eggs. In Brush-turkeys, older eggs are often stained brown from the moist earth that surrounds them, and also become more rough shelled as they age.

Table 5.1 describes the remarkable adaptations evident in mound-builder's eggs. One special adaptation is the extremely thin shell. By comparing the thickness of the shell with its relative size in the eggs of other galliform birds, David Booth calculated the 'expected' thickness of mound-builder egg shells. He found that mound-builder eggs were 31% thinner than the expected thickness.[3] Furthermore, the shell becomes progressively thinner during incubation, as the embryo gradually utilises calcium from the shell in the construction of its own bones. But why would birds risk breaking their thin-shelled eggs? The answer lies in the high humidity and extreme gas conditions inside the incubation mounds. For bird embryos to develop, water needs to be transferred from inside the egg to the outside while carbon dioxide, a by-product of the embryo's breathing, needs to be exchanged with fresh oxygen from outside the egg (through the process of diffusion). In mound-builder eggs, despite the high humidity and high carbon-dioxide content of the material surrounding the egg, gas exchange is greatly enhanced by the thin shell of the egg and the large size of the pores in the shell. Compared to other birds of similar size, the capacity of a mound-builder's eggshell to conduct water and gases is about twice as high[4].

Megapode eggs are remarkably large compared to the bird's body size. This can be explained by the fact that they need to contain a chick that is very advanced at the time of hatching (see Chapter 7) and that the embryo

**Table 5.1 Summary of special features of mound-builder eggs and embryos compared to other birds.**

| Mound-builders | Why? |
|---|---|
| Eggs are extremely large compared to female's body weight, exceeding all other birds except kiwis. | Eggs contain lots of yolk to sustain the embryo and enable the development of a chick that is at a far advanced stage (precocial) at hatching. |
| Eggs are not laid in discrete clutches as in other birds. | Eggs are very large and females can produce only one egg at a time. Eggs are deposited in mounds individually over many months and embryos develop and hatch separately. |
| Eggs are not turned during incubation, as in other birds. They remain in the same position, with the blunt end upwards, throughout incubation. | Females deposit them in the mound material and do not disturb them again. |
| Long incubation period | Prerequisite for the development of highly precocial chicks. |
| Eggs contain higher amounts of yolk than in most other birds. | To maintain the embryo throughout the prolonged incubation period; some yolk also serves as food reserve for the days after hatching.[5] |
| Egg shells are thinner than in most other birds | To facilitate the loss of water from the egg and the diffusion of oxygen and carbon dioxide to and from the egg.[22] |
| Eggs contain no fixed air bubble as in other birds. | Chick does not breathe air from an air cell before hatching. High humidity in the mound does not favour the formation of an air cell.[4] |
| Large variations in egg size within one species, greater than in most other birds. | Unknown, probably because the production of large eggs requires considerable energy investment and not all females have sufficient energy available to them. |
| Incubation temperature affects sex ratio, first evidence for birds. | Unknown, see discussion in the text. |

is incubated for a long period of time. Through the long period of incubation, enough yolk needs to be provided to support the growth of a large embryo. Also, the size of the egg needs to be sufficient for some yolk to be available as a crucial internal food reserve for the first days after hatching. The chick incorporates this extra yolk into its abdomen shortly before hatching. For these reasons, megapode eggs contain relatively large amounts of yolk. In

Brush-turkeys and Malleefowl, yolk make up 48–52 % of the weight of the egg content (no information is available for Scrubfowl, but in other *Megapodius* species yolk content is even higher at 63–69%).[5] Among all birds, only the kiwis from New Zealand have an equivalent yolk content (65%). In other precocial birds, for comparison, yolk weighs, on average, 37% of the egg, and in altricial birds, only 22%.[6]

To produce highly precocial chicks and provide enough yolk, mound-builder eggs are, therefore, of a considerable size. On average, Brush-turkey eggs weigh 180 g, Malleefowl eggs 173 g and Scrubfowl eggs 126 g, and the average egg length divided by the width is 1.59 for Brush-turkeys, 1.56 for Malleefowl, and 1.62 for Scrubfowl.[7] These measurements are most revealing when compared to the size of the birds. The relative egg weight, that is the weight of the egg in comparison to the weight of the female, is roughly 10% in Malleefowl and Brush-turkeys, but more than 20% in Scrubfowl. This means that the smallest of the mound-builders produces the largest egg in comparison to its own body weight. The relative egg weight is also very high in comparison to other birds with similar body weight. For example, birds of about the same weight as Brush-turkeys (1.8 kg) usually produce eggs that weigh approximately 60 g;[7] Brush-turkey females produce eggs that are three times heavier than expected. In addition, they produce more eggs per year than any other birds of similar size (see Chapter 6).

**Figure 5.1 Australian Brush-turkey eggs vary considerably in size. Both of these eggs were taken from the same mound.** Photo: Ann Göth.

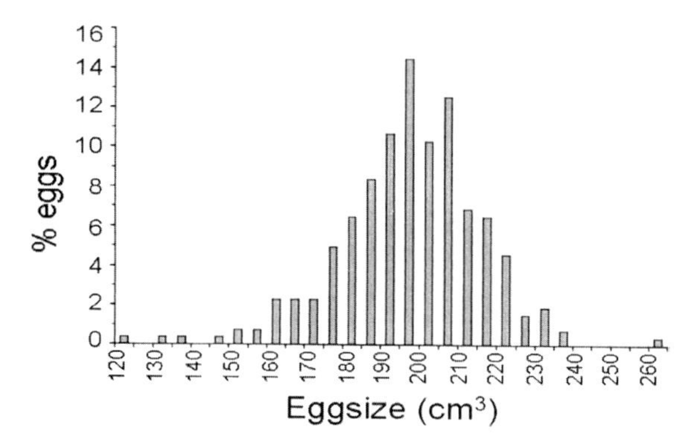


**Figure 5.2 Size of 263 Australian Brush-turkey eggs collected over two breeding seasons from 16 incubation mounds.** Reproduced from Göth (2007)[8] with permission from American Ornithologists' Union.

Another remarkable characteristic of mound-builder eggs is the huge variation in egg size within a species (Figure 5.1). Figure 5.2 shows that in Brush-turkeys, egg volume can vary from 123 to 262 $cm^3$, with an average of 197 $cm^3$, and with the largest egg weighing 2.1 times more than the smallest.[8] Very few other birds have such a high ratio of largest to smallest egg. A study of 39 other birds found that only 3% of them had ratios of 2.0 or more.[9] In Malleefowl, egg size also varies greatly, the smallest egg ever found in one population being only 99 mL in volume, while the largest was 200 mL.[10] In a different population of Malleefowl, egg size varied between 115 and 245 mL.[11] Chapter 7 discusses the effects of egg size on the chicks.

Finally, it is worth mentioning that megapode eggs are unique in lacking a fixed air bubble inside the shell. In other birds, this bubble is fixed in place at the upper end of the egg and enables the embryo to breathe air from the fixed air bubble shortly before hatching. This precedes the relatively gentle process of chipping away at the shell, which will eventually lead to hatching. Instead, the air bubble in megapode eggs is covered by the growing allantois (the inner egg membrane) and during the last few days before hatching, this structure becomes somewhat mobile. It is not used by the embryo, however, as breathing starts only when the membranes and shell are broken during the dramatic hatching process (see Chapter 7).

### *Clutch size*

Most birds lay their eggs at intervals of one to a few days, continuing until all the eggs to be produced are present in the nest. Only then does brooding start, with the bird's body providing the heat that starts the process of incubation and the steady development of the embryo. For most birds, this results in discrete clutches of eggs and, eventually, broods of hatchlings. Each embryo develops at the same rate and hatches at the same time. By starting incubation only when the clutch is complete these birds ensure that all of the hatchlings are of the same age and size. Megapodes, however, lay their eggs separately and at intervals of two to many days. As they produce eggs over a breeding season of many months, the chicks also hatch and emerge individually throughout this long period of time. They do not, therefore, produce a 'clutch' of eggs as do other birds but a series of eggs which develop independently. If one wants to compare 'clutch' size with other birds, it is best to count the total number of eggs laid by one female in one breeding season.

Normally, determining the number of eggs produced by a female mound-builder should simply be a matter of digging carefully through the incubation site and counting the eggshells found at the end of the breeding season. In Brush-turkeys, however, this task is made difficult, as females lay in several mounds and it is usually unknown how many females deposit their eggs within each mound within an area. So far, nobody has tried to follow individual females in the wild throughout the breeding season to find out how many eggs they lay. Estimates so far range from 18 to 24 eggs per season, based on the fact that individually marked females laid eggs in intervals of 2–5 days.[12] In captivity, the number of eggs produced by one female in one season ranged from 17[13] to 56.[14]

More reliable results on egg numbers have been obtained for the Malleefowl because usually only one female uses a single mound for egg laying. A long-term study over 13 breeding seasons showed that the average interval between subsequent eggs laid by one female was 6.4 days (n = 51 clutches), that egg numbers counted in individual mounds in different years varied anywhere between 1 and 28 eggs, and that the average total number of eggs per mound was 14.1 eggs (n = 124 mounds).[10] Other short-term Malleefowl studies have found similar results, with 14 (n = 34 mounds)[15] and 20 eggs per mound (n = 43 mounds),[16] and with one female laying a total of 34 eggs during a single season.[17]

Very little is known about egg numbers in the Scrubfowl. Early naturalists mentioned 'about nine' or 8–10 eggs, but these numbers seem too low compared to the other species, especially when females can lay their eggs at intervals of 9–20 days.[18] A Cairns-based study found that the interval between eggs averaged 13 days (range 9–20 days), but the authors were unable to count overall egg numbers.[18]

The topic of egg production is especially interesting when the weights of all eggs produced by one female during one season are compared with the female's body weight. With an annual production of 15–24 eggs, female Malleefowl produce 150–250% of their adult weight each year,[7] while a female Brush-turkey laying 28 eggs produces three times her own weight.[6]

When discussing egg numbers and laying intervals, we have to keep in mind that these can differ considerably between individual females and between years. It is most likely that egg numbers depend strongly on food availability, the fat or protein reserves of the female and the climatic conditions during the breeding season. The age of the female may also be influential. The 56 eggs produced by a single female Brush-turkey in captivity, for example, have very likely resulted from the bird's high quality diet. A long-term study of one Malleefowl population showed that the best predictor of clutch size was rainfall between May and December – the interval spanning both nest construction and egg laying.[10]

### *Hatching success*

The number of chicks hatching from an incubation mound does not always equal the number of eggs laid in it, as some eggs are predated or perish during incubation. Large goannas and feral pigs have been observed stealing eggs from both Brush-turkey mounds and Scrubfowl mounds, while introduced foxes are a major cause of mound destruction in Malleefowl.

Eggs can become addled because they are infertile or because the conditions, especially the temperature, are not suitable for their incubation. The adult birds may also damage the eggs when digging in the mound during laying or temperature testing, although this appears to be remarkably rare. Some embryos survive the incubation period, but die shortly before hatching for unknown reasons, while others hatch but are unable to dig themselves to the surface. However, despite the apparent challenge of digging through the mound materials, surprisingly few hatchlings perish during this time. One study of Brush-turkeys showed that of all eggs laid, 9.6 % perished due to addling, 3.6 % died before hatching, and 0.4 % died after hatching in the mound[12]; none were taken by predators. In comparison, a study of the fate of more than 1000 Malleefowl eggs found that 11.9% failed to hatch, 1.4% was broken by adults during mound work but 37.2% were predated by foxes.[10]

### *Incubation period*

Megapode embryos grow at a rate per day that is similar to that of other precocial bird species, yet the hatchlings are comparatively large and at a far advanced stage of development (see Chapter 7). To produce such well developed hatchlings, megapodes have evolved incubation periods that are much longer than in most other birds. Chickens, for example, hatch after approximately 21 days and Galahs *Eolophus roseicapilla* after 30 days. Megapodes, in contrast, have incubation periods between 49 and 70 days.

The length of the incubation period is strongly linked to the incubation temperature – at higher temperatures, embryos hatch sooner, while in cooler conditions, eggs take longer to hatch. The average incubation period for Brush-turkeys is 49 days, but this can vary by several days if eggs are incubated at significantly lower or higher temperatures.[7] When Malleefowl eggs are incubated in the laboratory at a strictly controlled temperature of 34°C, their incubation period is 62–64 days[7]. In the wild, their eggs take at least 60 days to hatch;[11] some eggs have hatched successfully after 78 and even 90 days.[11] No information exists on incubation periods for the Scrubfowl although this is 50–80 and 72–85 days for two other *Megapodius* species.[19]

### *Influences of incubation temperature*

In Chapter 4 we described the unusually high variations in incubation temperatures both within and between mounds. For mound-builders, these variations are a direct consequence of using mounds for incubation. Despite the birds' best efforts, differences in ambient temperatures, rainfall events and even the material in the mound can cause fluctuation in mound temperatures. Temperature fluctuations are usually not an issue for other birds as the brooding adult incubates the eggs at a stable body temperature. The only variation may be short-term cooling when the adult leaves the nest, but this is normally of little consequence.

Here, we report new findings on how these fluctuations in incubation temperatures of mound-builders affect many important features of embryos and chicks, including the incubation period, embryo mortality and the weight and sex of the chicks. These results, all from studies on Brush-turkeys, were largely generated by information provided by an aboriginal elder, Warren Canendo, who told one of us that after hot years, more female Brush-turkeys are seen, whereas after cold years, more males. His observations triggered a study in which freshly laid Brush-turkey eggs were artificially incubated at three different temperatures, all within the range of temperatures found in natural mounds (31°, 34° and 36°C). The result was that more males hatched at 31°C, more females at 36°C, whereas the sex ratio was almost exactly equal at 34°C.[20] This effect of temperature on sex ratios was statistically significant.

Results from this laboratory study were supported when incubation temperatures in natural mounds were linked to the sex of chicks hatching from these mounds. Eggs hatching females were found at significantly higher temperatures in the mounds (mean 33.7°C) than those from which males hatched (mean 32.9°C). Furthermore, sex ratios of chicks from individual mounds were significantly correlated with mean incubation temperature in those mounds – the higher the mean temperature, the more the sex ratio was biased towards females.[21]

This is, to our knowledge, the first time that an effect of temperature on offspring sex has been reported in birds. In reptiles, such effects are well known, and are due to the direct influence of temperature on the sex-determining genes. In birds, including the mound-builders, sex is determined by sex chromosomes long before incubation. The observed effect of temperature in the Brush-turkeys is, therefore, most likely to be based on a different mechanism to that of reptiles, although this has yet to be identified. One such mechanism could be sex-specific embryo mortality; in a laboratory study more embryos died at 31°C and 36°C compared to 34°C.[21] Possibly more female embryos died at the cooler temperature while more male embryos died at the higher temperature.

In addition, incubation temperature affects the incubation period and chick weight. An earlier study showed that at lower temperatures, Brush-turkey eggs take longer to hatch, while a recent laboratory study found that chicks incubated at lower temperatures weighed less than those from higher temperatures, although they were not smaller.[20] This suggests that lighter chicks had less internal yolk, and that much of this store of energy had been used up during the longer incubation period. Yolk is an important food resource after hatching, when chicks are still within the mound. During this time, they live on the yolk, which they had incorporated into their abdomen shortly before hatching. Having less yolk most likely increases chick mortality.

In summary, incubation temperatures affect the length of the incubation period, embryo mortality, as well as chick weight and sex. Given these effects, and the strong variations in temperatures between mounds, do females try to lay their eggs in the mounds with the best temperatures? A recent study indicates that they do: Mounds with mean temperatures ranging from approximately 32°C to 35°C received more eggs than cooler or warmer mounds.[8] Males thus have a higher breeding success if they manage to produce incubation mounds with the most suitable temperatures.

# 6
# GROWING UP WITHOUT PARENTAL CARE

*As a moth emerges from a chrysalis, dries its wings, and flies away, so the youthful Talegallus, when it leaves the egg, is sufficiently perfect to be able to act independently and produce its own food. (J. Gould 1842: 98)*[1]

*As for schooling, the chick neither knows nor needs parental care; schooling itself it instinctively adopts, as it grows to maturity, the life habits of its kind. To few animal folks has that power been given.*
*(A. Russell 1944: 85)*[2]

One of the inevitable tasks necessary for studying mound-builders is the excavation of mounds to search for eggs. We were often assisted in this laborious work by volunteers – who soon became dirty from top to toe – and sometimes we uncovered one or two chicks. Although we explained to our volunteers that this might happen, they were nevertheless astonished to encounter a fluffy chick deep within the soil and debris a metre below the surface. This surprise is easy to understand; no other birds start their lives having to battle through layers of soil. Furthermore, no other birds grow up without the care of parents, devoid of any guidance as to what to eat, what to be afraid of, where to go or whom to hang out with – with no parental role models from which to learn. Megapodes have among the most precocial young of all birds, with chicks that are remarkably developed at the point of hatching.

This chapter describes the minimal parental care provided to these chicks, as well as the behavioural and physiological adaptations of the hatchlings. It also shows how chicks survive, grow and develop the many skills they need for reaching adulthood, and asks whether some of these skills are innate or inborn, as opposed to learned from parents. All descriptions given here are of either Brush-turkeys or Malleefowl, as virtually nothing is known about the behaviour and development of Scrubfowl chicks. Table 6.1 lists some of the most remarkable features of mound-builder chicks.

## Appearance

Megapode hatchlings do not look much like their parents. Indeed, they more resemble the chicks of domestic chickens, or adult quail, with which they are often confused. Their feathers are brownish in colour, with the tail feathers reduced to very short woolly structures. Surprisingly for just-hatched chicks, most of the feathers on their wing (primaries) are almost fully grown, enabling the chicks to fly as soon as they have reached the surface of the mound. The plumage of all three species shows a variable amount of barring on the breast and shades of rufous or buff colours at different parts of the body, highly effective as camouflage. Even though the feathers on the chicks' bodies appear very woolly and resemble the dense and downy feathers of a domestic chick, their plumage is different in structure to that of other galliforms. (Technically, they do not have the semi-pennaceaous feathers of other galliforms but have fully pennaceaous feathers instead.)

## Hatching

While all other bird embryos break the eggshell with a sharp 'egg tooth' on their beak, megapodes use their legs (intertarsal joints) to dramatically escape the egg by pushing suddenly through the eggshell. Inside the egg, the head of the chick is positioned in between the legs, far from the eggshell. The thin eggshell and their strong legs enable the chicks to hatch feet first.

**Table 6.1 Summary of special features of mound-builder chicks, compared to other precocial birds.**

| Mound-builder chicks compared to other birds | Why? |
|---|---|
| Chicks hatch with their legs first, other birds hatch beak first. | This enables a more rapid escape from the eggs and the drainage of fluid from the lungs to enable breathing to occur fast. Also, chicks can trample down a cavity in the soil, in which they then emerge with the rest of the body. |
| Unlike other precocial birds, chicks hatch with fully developed wing feathers and pennaceaous downy feathers (compared to semi-pennaceaous in other birds). | Chicks need to be able to fly as soon as they have reached the surface of the mound to increase their chances of escaping from predators. |
| While other birds rest under their mother after hatching, megapode chicks rest in a cavity in the mound. | They need to fill their lungs with air, dry their plumage and lose the sheaths that enclose their feathers. |
| Chicks initiate breathing as soon as they have broken the eggshell. Other birds start to breathe from an internal air cell before hatching. | Eggs contain no air cell. Hatching is easy due to the thin shell and requires relatively little oxygen. After hatching, chicks can rest long enough to fill their lungs, and are fed by internal yolk. |
| Chicks have to dig themselves out, a process that is energetically more expensive than most activities performed by other bird hatchlings. | A result of the megapodes' unusual incubation of eggs in mounds of organic material. |
| After hatching, megapode chicks are more capable of thermoregulation than other birds. | This is facilitated by their relatively large body size, the ability to alter their metabolic rate and the good insulation properties of their well developed plumage. |

While kicking with their legs, the chicks could hurt themselves with their sharp claws. To prevent such accidents, their claws are enclosed in hard jelly-like transparent caps, which fall off soon after hatching. Also, their feathers are encased in sheaths, most likely to prevent them from getting muddy and dirty while they are still wet from hatching and stuck in the soil. After hatching, while resting in the mound, the chicks strip off these sheaths with their bill.

Megapode chicks also differ from other birds in that they do not start breathing air from an air cell before hatching. Other birds take some time to gently fill their lungs, slowly expanding these delicate structures as the fluids filling them is expelled. Roger Seymour's studies showed that instead, megapode hatchings rapidly fill their lungs immediately after emerging from the shell. Prior to this, the inner egg membrane (chorioallantois) was their life support and provided the oxygen. As soon as this membrane is ruptured, the chicks commence to breathe to fill their lungs.[3] Altogether it takes the chicks at least one hour to fully emerge from their egg.

Hatching with the feet first is of advantage in the soil, for two reasons. First, it enables the chicks to emerge from the egg very rapidly, and this again helps them to drain the fluid from their lungs as quickly as possible. As chicks do not start breathing air from an air cell before hatching, this draining of the fluid is an essential prerequisite for breathing. Second, the chicks can use their legs to trample down the soil and create a small cavity around their bodies before emerging with their head and starting to breathe.

## Behaviour within the incubation mound

When digging themselves out of their underground nest, megapode chicks are on their own. Their parents may only aid them inadvertently when digging holes in the mound for their daily maintenance activities, but otherwise they do not help them to the surface. In many other precocial birds, chicks synchronise their hatching by communicating in little squeaks with other clutch members while still in the egg. Before hatching, megapode chicks have no such contact with their siblings; the eggs are separated by dense soil, distributed throughout the mound and hatch completely asynchronously.

How do these chicks breathe when buried underground? Although the decomposition of the organic material leads to the production of large amounts of carbon dioxide, a dangerous gas in high concentrations, sufficient oxygen is also present to allow the chicks to survive. This is distributed within the mound by diffusion through spaces in the material, assisted by the adult birds when raking and mixing the mound material. These latter activities also loosen the material, prevent compaction and result in numerous small cavities throughout the mound, which serve as storage chambers for oxygen. Obviously, oxygen is essential for the chicks, especially when breathing heavily as they laboriously work to dig themselves out.

We observed the digging behaviour of chicks in a 'digging box' made of transparent Perspex positioned in a dark room (Figure 6.1). We watched 31

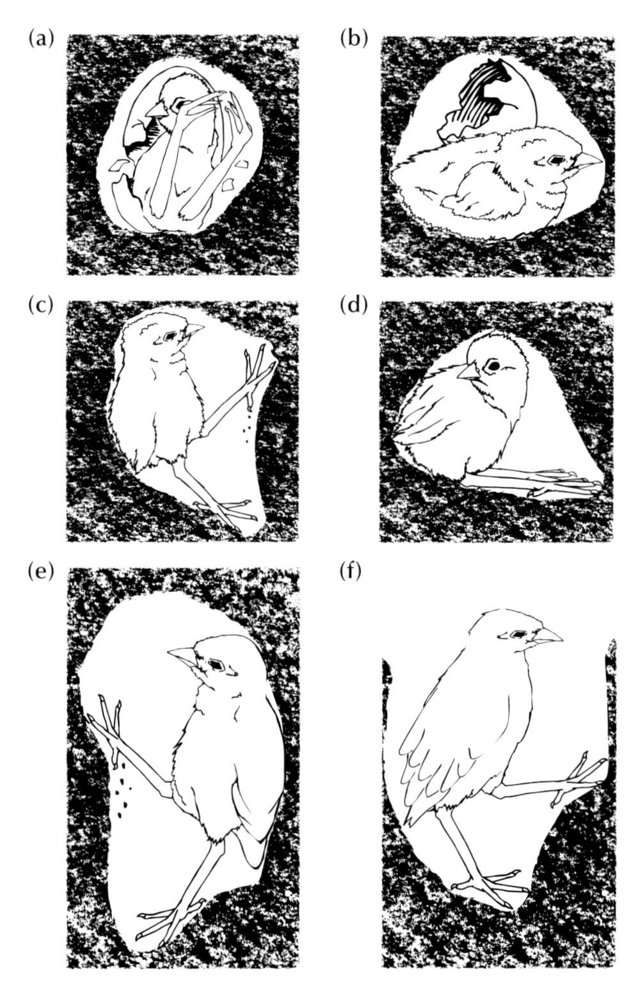


**Figure 6.1 Australian Brush-turkey chick during its first 40 hours after hatching.** (a) **Hatching with the legs first;** (b) **Long resting period after hatching in self-dug cavity, preening of the feathers;** (c) **Starting to dig upwards, eyes closed;** (d) **Long resting periods between periods of digging, trampling down of loose soil, resting on legs;** (e) **Intensive digging period during last hours, eyes closed;** (f) **emerging at the surface after an average of 40 hours, careful scanning of surroundings prior to running from the mound and hiding under cover.** Drawings: Ann Göth.

newly hatched Brush-turkeys that were carefully buried at the bottom of a 40 cm high column of heated soil.[4] In the wild, chicks hatch underground at a depth of between 40 and 150 cm, and our box resembled the conditions in a natural mound as closely as possible, though on a smaller scale and at the shallowest depth. It took these chicks an average of 40 hours to reach the surface, with some emerging after only 26 hours, while others took 55 hours to emerge. These results indicate that it takes baby Brush-turkeys at least a full day to dig their way to the surface, even in the best conditions.

After hatching, these hatchlings moved only enough to form a small cavity within the soil in which they remained almost motionless for an average of 16 hours. This fairly prolonged resting time was an essential preparation for the subsequent period of activity because it allowed the plumage to dry – important because digging with a wet plumage would cause them to get muddy and heavy. They also filled the many small branches of their lungs with air, and gradually increased their ability to regulate their body temperature.[5] Also during this time, the chicks lost the inner egg membrane that was still attached, as well as their feather sheaths.

After the initial resting period, the chicks began to dig themselves upwards through the mound material for an average of 21 hours, although this vigorous activity was in short bouts only, during which time they traversed only a few centimetres. During this time they almost always maintained a small cavity around themselves, in which they rested, preened and pecked at items in the soil, such as invertebrates. As they worked, they moved upward, alternated their activities between two types of movements. First, they pushed the top of their head and back against the ceiling of the cavity, causing soil to fall downwards. In the second movement, chicks sat on the lower part of one leg while scratching along the side and ceiling of the cavity with the other foot. They also trampled down the loose soil underneath themselves.

The greatest distance covered occurred within the last 1–2 hours before emerging at the top, with the chicks digging and pushing vigorously almost non-stop towards the surface. Just before they reached the surface, the chicks stopped digging and sat quietly for up to 5 minutes, as if listening to their surroundings. In the wild, this may help them to detect potential predators outside the mound. They then left the soil abruptly, flapping their wings and using their now very well exercised legs. In the wild, they would have run into the next thicket or cover to hide.

In summary, mound-builder chicks need to rest for many hours after hatching, just like other bird hatchlings. They do not fly or run as soon as they hatch, as it has been described in several books featuring these birds, because they are still deep beneath the surface of their incubation site. Furthermore, they are typically about two days old by the time they finally reach the surface. This clarification of the pre-emergence days is important for conservation programmes: chicks that are being released back into the wild should not be liberated until they have rested for sufficiently long periods of time and are about two days old.

While these observations of Brush-turkey chicks were made in an experimental 'digging box', they are nevertheless likely to resemble the natural behaviour of Malleefowl and Scrubfowl chicks. According to similar experiments conducted by Harry Frith using a glass-sided 'artificial mound', Malleefowl chicks take less time – from two to 15 hours – to dig themselves out through their friable sandy mound, compared to the matted sticks and plant detritus in a Brush-turkey mound. Overall, however, it appears that all mound-builder chicks need to rest for prolonged periods after hatching, for the reasons described earlier.

## The chicks' movements, environment and survival

As soon as they have reached the surface of the incubation mound, chicks can run very fast, and they need to do so to avoid predators and find a thicket or covered area in which to hide. They can also fly, as their flight feathers are already fully developed. Brush-turkey and Malleefowl chicks, however, do not usually fly during the first days. Instead, they either crouch or run away from predators. Scrubfowl chicks are better fliers and can fly directly from the ground into a tree when only about two days old.

Megapode chicks can cover remarkable distances during the first days after hatching. Radio-tracking studies showed that Malleefowl chicks moved, on average, 627 m per day (n = 19 chicks), and some of them averaged at least 2 km per day for one or more days after they left the incubation mound.[6] Brush-turkey chicks moved, on average, between 100 and 200 m during the first five days after reaching the surface (n = 71 chicks), with some travelling up to 800 m per day and others only 5 m.[7] Like the adults, Scrubfowl chicks are known to be extraordinary in their ability to disperse long distances, sometimes appearing on islands where these birds do not breed, such as Booby Island in the Torres Strait, and even landing on the decks of ships at sea.[8] This indicates that, they too, disperse far after hatching and even cross long stretches of open water.

Why do some chicks disperse so far and others stay in the same area? Most likely they are searching for a suitable environment in which to live. The radio-tracking studies showed that chicks unambiguously preferred to live in areas with dense ground cover. With Brush-turkeys, these were typically areas covered in introduced shrubs such as Lantana *Lantana camara*, Blackberry *Rubus fruticosus*, or other similar thickets, while the chicks avoided open rainforests with little ground cover.[7] In Malleefowl, chicks stayed in the dense ground cover provided in mature mallee habitat and few moved into recently burned mallee and open woodland habitats to search for grains and other food.[6] The Brush-turkey study also showed that chicks were less likely to be killed by predators in a thicket, whereas this likelihood was higher than expected in the open rainforest.[9] Clearly, thickets provide better protection from predators.

Survival in mound-builder chicks is often very low, especially in areas inhabited by feral cats or foxes. When Ann Göth and Uwe Vogel radio-tracked Brush-turkey chicks at two different locations, none of 43 chicks survived the first three weeks at a location with relatively little thicket cover, while 12% of 33 chicks survived the first three weeks at a second location with more thickets present.[14] Importantly, there was no evidence that the radio-tracking tags attached to the birds influenced this high mortality.[10] In a study of Malleefowl conducted by David Priddel and Robert Wheeler, none of 31 chicks survived for longer than 107 days and 71% were dead after 10 days,[11] and in Joe Benshemesh's study, not one of 21 chicks survived for more than 29 days.[6]

What are the causes of death? In most cases, introduced mammals are the main culprits. Cats accounted for the highest proportion of all deaths in Brush-turkey chicks (29–51%), followed by birds of prey (10–29%) with only a few chicks killed by foxes or dingoes or dying of unknown diseases.[9] Similarly, the radio-tracking study by Priddel and Wheler found that up to 68% of Malleefowl chicks were killed by introduced predators, principally foxes, and 39% by raptors.[11] In a recent study by Jessica van der Waag, on the other hand, introduced and native predators took a roughly equal toll on 20 Malleefowl chicks that were only a few days old when released: 40% were killed by introduced cats and foxes, the rest by raptors (30%), currawongs (10%), goannas (10%), an unknown predator (5%) or starvation (5%).

## Recognising food and predators

The chicks of all three mound-builders are omnivorous, feeding on a wide range of small invertebrates, seeds and fruit found on the ground. Yet, how do they know what to eat after hatching? Without parents to learn from, they receive no guidance when it comes to choosing an adequate diet from the many nutritious, non-nutritious and even toxic substances in their environment.

A study of Brush-turkey chicks focused on their ability to distinguish edible items, such as mealworms, fruit or seed, from non-palatable items, such as pebbles.[12] Two-day-old chicks, which had never encountered food before, were given a choice between a range of items. They clearly preferred the mealworms and pecked at the pebbles least often. The fruit and seeds were also eaten. Initial pecks at pebbles soon ceased, indicating that the chicks learned from trial and error. Most likely they identified fruit or seeds by their reflective surfaces and mealworms by their movement, and may also have used the contrast to the background as an identification cue. It appears that these chicks have the ability to innately recognise such aspects that many food items have in common, and this is then being fine-tuned by some trial and error learning.

A similar innate response was observed in a study on predator recognition.[13] Brush-turkey chicks were kept in a large outdoor aviary and a cat and dog were trained to walk from a hide through the aviary and back to the hide. In addition, a large natural-looking rubber snake was pulled through the aviary, and a silhouette of a bird of prey was flown overhead. Some chicks were only presented with playback of alarm calls uttered by local songbirds. The latter experiment was done to observe whether chicks recognise the warning calls of other bird species in lieu of parental calls, since Brush-turkey adults do not utter alarm calls of their own.

When the chicks heard the alarm calls of songbirds associated with the approach of a predator, they stopped what they were doing and looked around with their heads stretched upwards. Chicks of many other species react this way, but typically in response to alarm calls of their parents, rather than that of other birds. In the case of parentless Brush-turkey chicks however, the alarm calls of other birds have been exploited to aid survival.

Those chicks that were presented with either a cat, dog, snake or bird of prey responded to all these stimuli as well. Moreover, their response differed with the type of predator presented. They crouched to the ground when the raptor or cat approached, but mainly ran away when seeing the snake or dog. At the same time, however, they also showed the same types of anti-predator behaviour in response to other objects such as a moving cardboard box that was either pulled through the aviary or flown overhead, to provide a control for the behavioural experiment. These observations show that chicks may not recognise the predators per se, but respond to all approaching moving objects . Smaller objects evoke mainly crouching and larger objects mainly running. Such an innate response makes sense for Brush-turkeys: they evolved in the presence of a number of predators that differed in size, speed and behaviour, many of which have now become rare or extinct. With such a wide array of potential predators, it is better to take evasive action than to waste time on identification.

The observation of the chicks' response to a cat also revealed a disconcerting truth about their ability to cope with such introduced predators. Instead of running away, chicks crouch to the ground, relying on the camouflage. This makes them easy prey for cats and may explain why so many are killed by such predators.

## Species recognition and development of social behaviour

When moving around in the understorey, the chicks of mound-builders may occasionally meet up with other chicks, but they have never been observed following any adults. Malleefowl and Brush-turkey chicks are known to form loose social groups of two or three that stay together for a few hours or days before they separate again. Sometimes, all chicks in such groups are of the same age, at other times, younger chicks follow older ones. Most of the time, however, young mound-builders live a solitary life, without prolonged contact with others. In Ann Göth's radio-tracking study, free-ranging Brush-turkey chicks were detected with another chick in only 6% of all encounters (n = 166 encounters with a total of 31 chicks aged 2 days to 4 weeks).[14]

For other birds, meeting siblings, parents or other conspecifics greatly affects the development of their social behaviour. Young chickens, for example, do not show all the typical behaviours of adults, such as aggressive displays, and only gradually develop such social behaviours. This serves to avoid aggression, because young birds that behave like adults might be regarded as rivals by territorial adults and could be harmed. Mound-builder chicks, in contrast, do not form groups with adults and are thus not at risk of offending them. This may explain why all of the social behaviours – except mating behaviour – normally found in adults are also observed in chicks.[15] Hence, social behaviour also develops without social experience and requires no learning from conspecifics.

Another question to be asked about mound-builder chicks is how they recognise conspecifics. It is well known that in most other birds, such species recognition is based on a process called 'imprinting', in which young birds form attachments to objects that they are exposed to during a particular 'sensitive period'. Usually, this sensitive period occurs briefly after hatching and the object is a parent. If, however, the object happens to be a particular person, young geese or other birds may follow that person instead of their own parents.

Mound-builder chicks are unlikely to imprint on their parents or siblings as they do not reliably encounter them during any equivalent 'sensitive period' early in their lives. And indeed, one experiment with chickens and Brush-turkeys showed that while the chickens immediately followed a moving ball after hatching, Brush-turkeys showed no inclination to do so at all.[16] So what cues do these chicks use to recognise members of their own species? Do they use calls of other chicks as a recognition cue, or perhaps specific behaviours or plumage colours? These questions were asked in a series of 'two-way' choice tests, in which Brush-turkey chicks were placed in the middle of a large aviary where they could run into two different 'choice arms' of the aviary. A stimulus was placed at the end of one choice arm, a control for the stimulus at the end of the other arm. Sometimes, the stimulus was a loudspeaker that played calls of other chicks, and the control was a silent loudspeaker. In another case, the stimulus was a stuffed Brush-turkey chick that contained an engine for remote-controlled cars, which made the chick move up and down as if pecking at the ground (the so-called pecking robot), and the control was a robot chick that moved its head from side to side (the scanning robot). In a final test, the stimulus was a stuffed robotic chick that appeared in a different colour (due to coloured filters mounted above it), and the control a plain-coloured robot chick.

These tests showed that the pecking behaviour of another chick clearly attracts chicks to each other. The chicks approached the pecking robot more often than the scanning robot, and they often started pecking at the ground as soon as they saw the pecking robot do so.[17] Obviously the chicks have evolved a high sensitivity to a movement pattern that is functionally important, as pecking likely indicates a potential food source. In the wild, however, other species may also peck at the ground and it would not necessarily be sensible for young mound-builders to regard these as conspecifics. Our tests showed that the young Brush-turkey chicks also used a second set of characteristics: certain body regions that reflect the light at short wavelengths. When we mounted a filter for UV light or short wavelengths above the robot chick, the chicks approached it significantly less than when the model was in normal light. When we mounted filters for medium or long wavelengths, however, there was no difference in which type of model was approached.[17]

A specific apparatus designed to measure reflectance also revealed that the chicks' legs and beak reflect strongly in the UV light. It might thus be that chicks use the reflectance of these body parts as an additional species recognition cue. In adult Brush-turkeys, the male's large neck wattle has also been found to reflect in UV light (see Chapter 3) and in this case, it is suspected that this may have a role in male attraction.

Interestingly, the response of these Brush-turkey chicks to the playback of chick calls differed to the visual stimuli in that the chicks did not approach the loudspeakers. They did, however, lift their head and scanned their surroundings when they heard the calls of other Brush-turkey chicks, suggesting that they may have recognised these calls.[18] In summary, this series of tests on species recognition showed that Brush-turkey chicks hatch with an innate perceptual bias for relatively specific cues which enables them to recognise conspecifics when they first encounter them.

## Growth and reaching adulthood

Brush-turkey chicks grow amazingly quickly. They start life looking similar to quail, but at only four weeks they resemble miniature adults, although they lack the vivid colours on the head. By nine months, they are barely distinguishable from adults. They start replacing their original brown feathers with black feathers by the age of three weeks, and their tail feathers first appear when they are two weeks old. At about this time they also start losing the feathers on their head and neck, which exposes the red and yellow skin underneath.

Figure 6.2 shows how fast Brush-turkey chicks gain weight after hatching, especially during the first four months. However, this process only starts after their first week of life; during the first days, the weight of these hatchlings initially drops as they use up their internal yolk reserves. Males and females are of similar weight at hatching, but by about five months, males are notably heavier than females, and this difference continues into adulthood. At the age of 10 months both males and females have reached approximate adult body mass.

Malleefowl, on the other hand, seem to grow and gain weight more slowly than Brush-turkeys (Figure 6.2). Eight chicks raised in captivity by David Booth only reached 75% of their adult body mass by the age of 1.5 years.[19]

The exact age at which young mound-builders are first able to breed is not known for any of the three species. In Brush-turkeys, males may start building mounds as early as 4.5 months of age,[20] whereas anecdotal observations from the early 20th century suggest that the youngest age at which females start egg laying is 11–12 months. The onset of mound building in males is not, however, equivalent to sexual activity. Darryl Jones observed six one-year-old males in the wild, and of these, four constructed mounds during their first year, but all were expelled by older

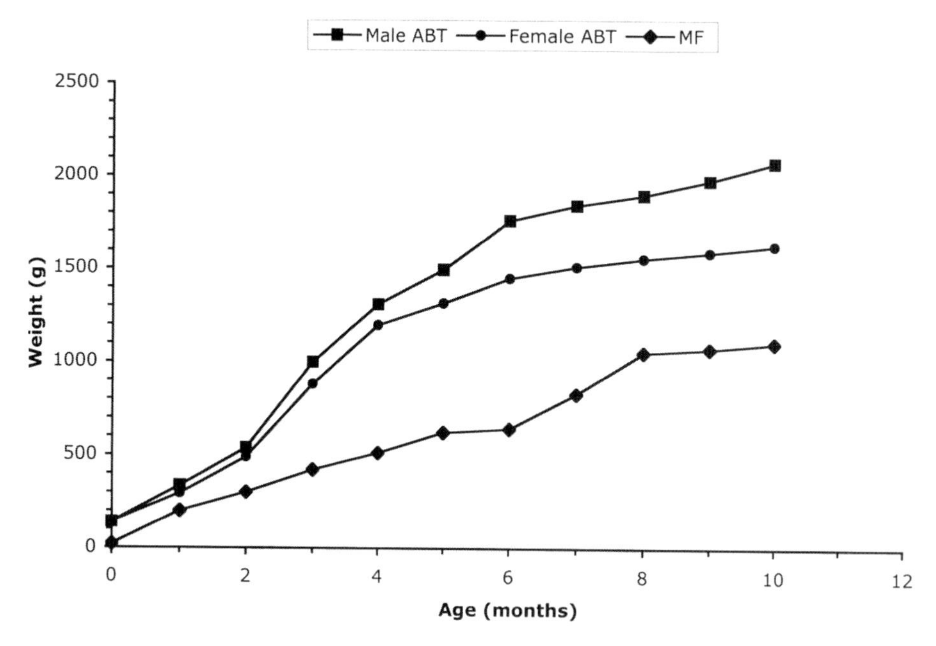


**Figure 6.2 Weight gain of male and female Australian Brush-turkeys (ABT) and Malleefowl (MF) during the first 10 months after hatching. ABT values are means for 11 males and 12 females, MF values are means for eight birds of both sexes, all raised in captivity.** Data for ABT from A. Göth, for MF from David Booth (used with permission).

males and presumable did not mate with females. In the second year, only one of these males was successful in building and keeping a mound, mating with females and receiving eggs.[21]

Observations by David Priddel, Robert Wheeler and Harry Frith suggest that Malleefowl may not start breeding before they are about four years old.[22,23] Younger males of two or three years of age do often start working mounds, but it appears that they do not usually breed at this age. It may well be that, as in Brush-turkeys, young Malleefowl males start practicing their mound building skills early but need to learn how to construct mounds that females accept for egg laying.

# 7

# SOCIAL AND REPRODUCTIVE BEHAVIOUR

*With the constant need to tend and guard the mound, the mallee-fowl has little time for family life. He and his mate lead more or less separate lives; the male worries about the mound and the female, burdened with egg-production, concentrates on collecting food for herself. They meet frequently at the mound and occasionally in the bush, but on the whole live lives of solitude. (H. Frith 1962: 63)*[1]

## Males, females and mating systems

In *The Mallee-Fowl*, Harry Frith summarises his years of intensive study in an evocative portrait of life for the Malleefowl in their harsh and unforgiving environment. Having shared the heat, toil and loneliness of the mallee with his subjects, he can speak with some authority of their habits and demeanour and quotes an early naturalist, K. H. Bennett, with approval: *'Its actions are suggestive of melancholy, for it has none of the liveliness that characterises almost all other birds, but stalks along in a solemn manner as if the dreary nature of its surroundings and its solitary life weighed heavily on its spirits'.*[2]

As well as trying to convey an accurate impression, Frith also provides an effective description of the profound influence the environment has on the daily lives of these birds. The tough conditions Malleefowl must face – which contrast greatly with other megapodes living in moist forests – not only impact on their individual work schedules, but also force them to spend much of their time apart, engaged in very separate tasks. Nonetheless, Frith also points out these birds live in pairs, with lifelong bonds, and despite their largely solitary lives, are distinctly monogamous.

Frith's assumption that his birds were monogamous was based on accumulated observations over a long time – he only rarely saw unpaired birds and all interactions between strangers were aggressive. But he also employed one of the long held categories used for describing the reproduction arrangements – 'mating systems' – formed between males and females of all sexual species. *Monogamy* refers to exclusive relationships between partners of each sex for at least one breeding season, but often for many years, and sometime even for life. More than 90% of birds raise their chicks in typical one male, one female partnerships. The main alternative among birds is *polygyny*, which often includes a 'harem' in which a single male sequesters a collection of females. Interestingly, polygyny is the normal arrangement for almost all the other galliforms, typified by the farmyard rooster and his well guarded group of females, but also by the quail and pheasants. Least common is its opposite, *polyandry*, in which a single female mates with a number of males. Although rare among birds, polyandry is well known in Australia because it is practiced by our biggest birds, emus and cassowaries. And finally, if there appears to be no obvious patterns to the sexual relationships between the sexes, the unfortunate term *promiscuous* is usually used, though the implication of a lack of selectivity is almost never appropriate.

Although assigning these basic mating system classifications to a particular species may appear to be straightforward, closer examination of the actual relationships between males, females and their young have shown that even the simplest arrangements can be more complicated than expected. Traditionally, mating system classifications have been based on careful observations, often, where possible, of individually marked birds. However, even the most diligent naturalists cannot watch their subjects all of the time, and there was always the possibility of secret liaisons. Recently, developments in molecular biology have added dramatically to our ability to understand the reproductive relationships between individuals, especially in terms of parentage. Perhaps the most astonishing general finding to have emerged from these genetic studies is the discovery that most of the species we have traditionally regarded as monogamous are actually far less faithful than ever suspected. Although studies continue to show that males and females remain together to raise the chicks in their nest, parentage comparisons show that between 5 to 40% of these young have been fathered by a male other than the female's mate. Today, finding a species in which extra-pair matings do not occur is proving to be a major challenge! (Among Australian species, silvereyes and robins appear to be among the few that appear to be faithfully monogamous.)

However, rather than view such things moralistically (as did many earlier commentators who spoke of 'the male and his wife' and 'adultery'), it is important to ask why these activities are so common. Modern evolutionary approaches to animal behaviour attempt to understand the possible benefits that a particular behaviour or trait may have for an individual. For example, a female bird may select to remain with a particular male because he is a good parent or provides a safe nest site yet chooses a neighbouring male to father her young because he may provide better genes, as evidenced by his superior plumage or song. Although this scenario may seem far-fetched, such findings are becoming increasingly common. Of course, discerning this level of detail requires both careful observations in the field and sophisticated genetic techniques.

These insights and approaches have led to a revision of the traditional meanings of the mating system categories. Monogamy, in particular, while clearly being associated with pro-longed pair-bonds between individual males and females, cannot be assumed to mean sexual exclusivity. Until proven otherwise, monogamy is best understood as meaning 'social monogamy'. This background is essential to our discussion of the reproductive behaviour of the mound-builders. But, as in every aspect of the lives and behaviour of these birds, first we must consider the influence of their method of incubation.

## Parental care, before and after hatching

As we have made clear throughout this book, the use of external sources of heat rather than the warmth of the parent's body has lead to a cascade of adaptations in many aspects of these birds' lives. In particular, eggs and embryos are extraordinarily resilient and well adapted to being incubated in mounds of organic material (Chapter 5), and chicks are highly precocial, leading an independent life without forming any bonds with their parents (Chapter 6).

From the perspective of the adults, these adaptations mean complete liberation from what, for most animals, is one of the most demanding and risky aspects of their lives: ensuring that as many offspring as possible reach independence. Normally this would include a wide range of tasks including brooding, sheltering from weather, protection from predators, finding food and facilitating the development of social and survival skills. In most birds these typical parental tasks occupy a vast amount of time and energy; for megapodes the effort expended in these parental investments can be directed elsewhere, to other means of enhancing reproduction. This comes down to two main tasks: provisioning the egg with the optimal materials required for the best development of the embryo; and providing ideal incubation conditions. While the first of these is obviously limited to females, both males and females may be involved in working on the incubation mound.

## Influences on megapode mating systems

### *Australian Brush-turkeys*

Prior to the first intensive studies on this species, all megapodes were assumed to be strictly monogamous. Although there were numerous reports of more than one female laying in a mound, there was still no serious suggestion that this was anything but mound sharing by other pairs. The first confirmation that Brush-turkeys were non-monogamous emerged from an afternoon's observations of the interactions between a male and the succession of females who visited, mated and laid in his mound.[3] This was the decisive moment in 1979 that lead to Darryl Jones's life-long obsession with these birds. Since then, additional studies by Sharon Birks and Ann Göth have added greatly to what is known about the complex social and sexual lives of these birds.

For male Brush-turkeys, the single most important means by which they may influence their potential reproductive fortunes is through the possession of a mound. Only by having a mound can males hope to attract and mate with females; access to a mound is the only way females can get their eggs incubated. This species has opted for a straightforward gender division: females concentrate on egg production, males on mounds.

It would be no exaggeration to say that male Brush-turkeys are obsessed with, and possessive about, their mounds. They start by carefully selecting the location of a mound site in relation to the surroundings, the amount of sunlight and apparently the quality and quantity of suitable leaf litter available within the immediate vicinity.[4] Construction requires sustained and rather conspicuous activity as the male begins to gather soil, leaves and humus from a broad area of the forest floor. This inevitably draws the attention of other Brush-turkeys, including males who may wish to challenge the presence of this individual in this particular location, and females who will take a prolonged interest in the male and his work. Young and otherwise subordinate males that attempt to construct a mound in an area already occupied are likely to find themselves rapidly excluded, their mound taken over or simply abandoned. In one study,[5] over half of all mounds constructed were abandoned before receiving eggs, largely because of expulsion by other males. Hence, to succeed in attracting females, a male must be able to defend his mound from other males.

He must also be able to keep it functioning for extended periods of time. Brush-turkey eggs require about 49 days of incubation and with eggs being laid over a period of up to eight months, a mound will require considerable regular work to function as a viable incubator for this length of time. Although the initial construction work is certainly the most intensive – the male must move 2–4 tonnes of soil and leaf litter into the pile – the relentless decomposition of the organic matter within necessitates continuous 'feeding' of the micro-organisms.[6] Without such additions of fresh material, the mound heat production will eventually decline, although this appears to be a slow and drawn out process. Fully operational mounds, which had been abandoned or where the male had died, continue to function adequately – maintaining temperatures of 32–34°C – for more than six weeks before starting to cool.[6] However, the longevity of a mound may be constrained by features of the site such as insufficient leaf litter to add to the mound. One way for males to prolong their possession of an incubator is to construct a second mound, which may be operated simultaneously with the original, or as a replacement for the first.[5]

The construction of a second mound effectively doubles the energetic costs for the male, even though the benefits associated with attracting females may be enhanced. An even more effective way of acquiring a second mound would be to avoid the costs of construction altogether by getting another male to do the work. This devious approach appeared to be employed by a number of dominant males in a study that investigated the tactics males used in relation to mounds.[6] Instead of quickly expelling subordinate males that started mound building relatively close to their own mounds, these males were tolerated until most of the hard work had been completed; the constructors were then expelled and their mound usurped.

All of this competition and effort on the part of males has a single and obvious goal: to attract females. Although we now know that females of many species select carefully among a series of potential mates, most will mate only once with a single or possibly a couple of males. For females, these few copulations are all that is required to fertilise their entire reproductive output for the year; making sure that the choice of mate is a good one, therefore, is vitally important. In contrast, female Brush-turkeys seem to have an opportunity to exercise an extraordinary level of selection and assessment of prospective males.[7] It is useful to think of female megapodes as laying many separate eggs – each potentially fertilised during a separate copulation – rather than a clutch of eggs all fertilised by one mating. This certainly seems to be the case with the Brush-turkey: with males preoccupied with the defence of their mounds, females are seemingly free to wander among the entire array of potential males without restriction, throughout the prolonged period that mounds are maintained as incubators. This allows females the chance to check and compare all of the males and their mounds.

Given this unparalleled level of unrestricted mate choice, what might females be interested in, in their potential males? Theoretically, there are three main issues for females assessing a mate: (1) physical and behavioural characteristics of males that reflect their genetic qualities; (2) the ability to provide suitable incubation conditions for eggs; and (3) other resources associated with the males such as food or shelter. As female Brush-turkeys spend relatively little time with the male or near his mound, the third of these issues can be discarded. This leaves morphological features of the male, and his ability to incubate eggs, as the most likely criteria being compared by females.

The first studies by Darryl Jones found that males that provided incubation sites for the longest periods received the most eggs, and that having two mounds meant twice as many eggs.[7,8] Male physical and behavioural qualities appeared of little importance compared to mound features. This initial impression – that females are more influenced by aspects of incubation than males – has been strongly endorsed by several recent studies. First, detailed studies of parentage by Sharon Birks found that females typically laid a series of eggs in the mound of one male, before switching to another male and his mound.[9] Thus, there was little loyalty to specific males within and between breeding seasons; indeed, 28% of chicks were found to have been fathered by males other than the one tending the mound from which they hatched (in the most extreme case, 44% of eggs in one mound were not the mound owner's[9]). Second, females did not stop laying in a particular mound even when the original male was replaced by another male, who had previously been unable to attract females to his former mound.[10]

More evidence comes from Ann Göth's recent work on the relationship between egg size and mound temperature in this species.[11] These studies showed that mounds differed significantly in average incubation temperatures, and that females preferred to lay in mounds that were within the 32.5–34.5°C range. Mounds with these incubation temperatures received more eggs, which tended to be larger than average. As explained earlier, bigger eggs are more likely to hatch bigger chicks that grow faster and consequently have a greater chance of survival than smaller eggs (Chapter 6).

However, it would be too simplistic to suggest that females are only influenced by mound characteristics, ignoring completely the looks and behaviour of the males. The fact that male Brush-turkeys have evolved notable features such as the booming vocalisation, and the conspicuous and adjustable neck wattle (which is highly visible in UV light), for example, suggests a more subtle role in mate choice or advertisement; future investigations on these key features are needed.

Perhaps the most impressive male Brush-turkey behaviour – unique among megapodes – that is unequivocally directed towards females is the peculiar 'prostrate' display. Always performed by a mound-possessing male in the presence of a visiting female, the male lowers his body to the ground with tail and wings flattened and spread laterally over the substrate, while pecking repeatedly at the ground but not foraging. This display is given both on the mound and away from the mound in the vicinity of the female. It has been reported in captivity,[12] among high density suburban populations[13] and among wild birds.[8] Given the social context, the most likely functional explanation for this display is that it is carried out to stimulate a poorly motivated female, or to attract her to the mound.

### *Malleefowl*

From the tropical exuberance and sexual variability of the Brush-turkey, consideration of the apparently austere social life of the Malleefowl provides a stark contrast. Although the Malleefowl may be safely regarded as monogamous, in contrast to the Brush-turkey, there are nonetheless some important similarities between the two species, especially in terms of mound attendance by the males and frequent separations between the sexes. In general, the male and female Malleefowl appear to be stolidly occupied with their respective priorities: the constant effort required to maintain a functioning mound by the male; and the necessity of foraging for sufficient food for egg production in the female. As described by Frith and others, the harsh and demanding environment in which they live is surely the overwhelming factor influencing the social lives of these birds. While the male is obliged to stay near the mound to gather the tough organic matter required to fuel the decomposition within, the female must wander far afield in search of the nutrition necessary to produce many large eggs.

While this general picture of Malleefowl life – based primarily on Frith's studies near Griffith – remains sound, more recent studies in other places have found important differences in the social behaviour of the species. For example, while the Griffith birds appeared to spend relatively little time together at the mound, Malleefowl studied in the Murray River Basin of South Australia often worked together on mounds for relatively long periods of the time.[14] This cooperation was especially evident around the times that eggs were being laid, a process in Malleefowl that requires the removal of at least 850 kg of soil and sand to gain access to the incubation area of the mound. During this laborious activity – lasting an average of 518 minutes – the pair worked at almost identical rates, itself an unexpected finding as the female carried a large egg that was ready to be laid. Frith's males, in contrast, were solely responsible for the huge effort of opening the mound for laying.[15] The Murray River birds also spent relatively more time together when not working on the mound and were distinctly more social than the Griffith pairs.[16]

Despite these regional differences, all Malleefowl form distinct pair-bonds and exhibit a range of behaviours associated with well maintained relationships. For example, there are no apparent pre-copulatory displays or behaviours, and mating may occur at times unrelated to fertilisation. Furthermore, during reunions, pair members perform a closely coordinated greeting display, with the birds circling one another, ruffling their feathers and spreading their wings.[17] Malleefowl also employ various vocalisations as contact calls between individuals, sometimes including a duet performed by both male and female together. It should be noted, however, that both the greeting display and duet have only been recorded in captive Malleefowl[16,17] and not often described for wild birds.

Although Malleefowl appear to be distinctly if not strictly monogamous, it is now well recognised that individuals of almost all species will take advantage of an opportunity to improve their reproductive output. In the case of the Malleefowl, such opportunities appear to be remote and exceedingly unlikely: the density of birds is very low, mound and neighbouring birds are mainly well separated and most interactions with strangers, male and female, especially during the breeding season, are highly aggressive. There is, however, one definite case of a male Malleefowl forming bonds with two females.[18] In this case, the male managed to maintain two mounds some distance apart and divided his time almost equally between the two. The two females laid relatively large numbers of eggs – 30 and 29 (the average is usually 13–20 for Malleefowl) – in each mound. Although this confirms that even the Malleefowl can be polygynous when the opportunity arises, the circumstances in which this occurred were not typical for the species. First, significantly more rain had fallen in the area, facilitating the production of both food and damp organic matter for mounds. Second, these birds lived close to a picnic area in a National Park well used by the public and it is likely that the site provide much more mound material and supplementary foods than would be normal for Malleefowl.[18]

### *Orange-footed Scrubfowl*

Having briefly summarised the extensive information available on the two most well studied species, discussing the social and reproductive behaviour of the Scrubfowl limits us to only a few studies. Although this species is the most widely distributed of all megapodes, information from Australia is limited to two intensive studies of small numbers of birds observed near large urban centres, Cairns, north Queensland and Darwin, Northern Territory. Additional information comes from observations of the species in Indonesia, and relevant studies of other *Megapodius* species.

Whereas Malleefowl remain monogamous despite the pair members spending considerable time apart, we can be unequivocally certain of social monogamy in Scrubfowl because the male and female are almost never apart. Rather than males remaining close to their mounds throughout the breeding season and hoping that females will visit them, as is the case in both of the other Australian mound-builder species, male Scrubfowl stay close to their mates rather than to their mounds. One possible explanation for this different type of pair-bond seems to be that males may help their females to produce their large eggs, both by guarding them and defending a feeding territory. The large size of the eggs in *Megapodius* species (compared to the female's body size) has been suggested as a driving factor behind monogamy in another, closely related species.[19] As pointed out in Chapter 5, the Scrubfowl has the highest relative egg weight of all three Australian mound-builders. While producing eggs, females require a high intake of protein-rich food. They seem to benefit from a monogamous pair-bond in the form of a feeding territory, defended by the male, and the invertebrates he uncovers while feeding close to her. Also, while being guarded by the male, they can spend more time searching for food and are protected from forced copulations from other males. However, further studies are needed to confirm this hypothesis.

Scrubfowl pairs may be seen near or on their mound at any time of the year but spend progressively more time working together on the mound in the latter half of the year. In Darwin, Scrubfowl mound work was positively correlated to recent rainfall, with the birds being particularly busy following heavy monsoonal rains.[20] Unlike the Brush-turkey and Malleefowl, Scrubfowl tend their mounds much less intensively, sometimes not visiting for many days, but are more likely to work on the mound at any time of the year. Moreover, male and females almost always labour together on the different components of mound work such as gathering leaf litter, mixing, temperature testing and excavating holes for egg laying.[20] A quantitative comparison of the times spent in these activities revealed, however, that males performed significantly more of this work than did females.[21]

Scrubfowl are by far the noisiest of all megapodes, producing a wide variety of loud and complex calls. While detailed studies of the vocalisations and their function have yet to be undertaken, it appears clear that they are employed primarily to advertise the location of a pair. Although this may be regarded as being territorial in nature, there is actually little evidence that the defence of space is the main purpose of these calls. In possibly the loudest and most complex of the calls, both pair members engage in a prolonged duet. These duets are either produced throughout the day, or at night from favourite roosting sites in the rainforest canopy and are answered by other pairs. They most likely serve to indicate the location and presence of the pairs. In the Cairns study,[21] these vocalisations appeared to allow the numerous pairs living locally to share the area without confrontations; aggressive interactions only occurred when single birds were encountered.

In Cairns, these vocalisations also facilitated the sharing of access to incubation mounds; pairs approaching a mound would call and the pair present would leave without incident. Five of the 28 mounds in the study area were used by two pairs, although one pair undertook most of the mound work. Also, mound ownership changed regularly for those many mounds which were used continuously or even annually for decades. However, understanding these changes in movements and mound use requires the marking of individuals and this has only been undertaken on a few occasions. In the Darwin study,[20] for example, a banded male was found to be maintaining two mounds simultaneously. However, further details of this bird's social relationships with other birds was limited because the other birds visiting were not marked. Clearly, we need to know a lot more about the social behaviour of this fascinating species.

# 8 CONSERVATION AND MANAGEMENT OF AUSTRALIAN MOUND-BUILDERS

*All three birds are 'good eating' and when it is remembered that foxes and feral cats assist in the slaughter of malleefowl and that the large eggs weigh about 3 times as much as an average domestic fowl egg and are freely taken by settlers, the wonder is that the species have survived to even a limited extent. (A. Chisholm 1943: 49)*[1]

*Brush-turkeys have many human enemies: the birds are shot for food by settlers, in some districts, and their mounds are robbed, since the large white eggs are good eating. The Lowan is closely associated with mallee, and, deprived of its shelter and seclusion, soon disappears from age-old haunts of its species. (C. Barrett 1931: 115)*[2]

Megapodes have always been hunted for food, and their eggs collected for human consumption. In this chapter, we describe the impacts of egg collection and hunting on the three Australian species. We then discuss the conservation of each species separately, with particular emphasis on the Malleefowl, which has experienced a much more drastic reduction in its range and population numbers than the other two species. Malleefowl are now listed as Vulnerable under the Commonwealth legislation (*Environment Protection and Biodiversity Conservation Act 1999*), whereas both Brush-turkey or Scrubfowl remain secure or 'Of Least Concern', throughout most of their ranges. This chapter also discusses the reasons for the different fortunes of the three species and the challenges faced by land managers and other people involved in their conservation. We also include the rather unusual issues associated with 'urban mound-builders'.

## Egg harvesting

Eggs of mound-builders have always been a valuable resource for indigenous people, both due to the large numbers of eggs these birds produce, their large size, and the fact that they can be obtained relatively easily by digging up incubation mounds. In Australia, all three species were widely utilised by Aborigines. In some mallee areas, for example, Aboriginals camped annually in locations with Malleefowl mounds for the purpose of collecting and eating the eggs as soon as they were laid. However, apart from the possible extirpation of Scrubfowl from some tiny islands off northern Australia, indigenous use appears not to have greatly affected distributions. Today, the impacts of egg collection is only minor compared to the other threats faced by the mound-builders, and compared to other areas outside Australia, where megapode eggs are often consumed to such an extent that populations are in serious decline.

The first Australian settlers also collected eggs quite frequently, as described in many accounts from the mid and early 20th century.[1,2] A publication from 1954, on the 'palatability of bird eggs', describes those of each of the mound-builders as 'excellent eating' and also mentions that the nesting mounds of both Scrubfowl and Malleefowl were often 'ruthlessly destroyed' by white settlers. Another publication from 1909 mentions the value of Malleefowl eggs to travellers: '*Many a thirsty wayfarer whose water-bag has become dry, and who has been lost in these waterless areas, owes his preservation to the reviving influences of the sucking of Mallee-Fowl eggs*'. Today, egg removal appears to be much less common as it is illegal and also far easier to buy a carton of chicken eggs from a supermarket than to dig up an incubation mound.

## Hunting

Mound-builders were actively hunted from the start of European settlement. Clearly resembling other game species, the large bodies of both Malleefowl and Brush-turkey were especially sought after for the pot. The apparent impact of this spontaneous hunting was noted by visiting naturalists during the late 1800s with numerous reports of once abundant birds becoming increasingly scarce in locations near settlements. When laws protecting wildlife were first introduced in the 19th century, the mound-builders were assigned to the group of protected game birds, though for regulating their hunting. Full legal protection was then only granted from the early 20th century. The following describes the legal situation in NSW,[3] as a typical example of this pattern.

In NSW, Malleefowl and Brush-turkeys were among the first birds to be specified by colonial law, as early as 1866. This seems ironic, given the dire situation of the Malleefowl today. However, this early legal status was introduced specifically to preserve both species for hunting purposes. Under the *Game Protection Act 1866*, both Malleefowl and Brush-turkeys were assigned to a group of popular game species in the state and the act acknowledged the need to protect the opportunity for people to hunt them for sport. Nevertheless, both species may have benefited from this law to at least some extent, as it aimed to control illegal hunting by squatters throughout the state, which had become very common by that time. Furthermore, all species listed under this law were banned from being hunted for an initial five years following the introduction of the Act, and could then only be hunted during an 'open' season declared by the authorities, which largely corresponded with the non-breeding season.

The *Game Protection Act 1866* was followed by other legislations, such as the *Birds Protection Act 1881* and *1893*, under which both the Brush-turkey and Malleefowl continued to be listed as game birds for hunting. This primarily reflects the hunting value the community placed upon these species. It also highlights that for many decades, direct killing was the only acknowledged threat faced by mound-builders, whereas habitat destruction and predation by introduced feral animals was not recognised. This attitude slowly changed with the introduction of the *Birds and Animals Protection Act 1918* which, for the first time, protected all native species,

**Figure 8.1 Malleefowl hunting party in the Cobar district of New South Wales.**
Reproduced from Priddel and Wheeler (1999); photograph courtesy of Cobar Museum; source unknown.

except for 45 native birds and mammals listed as species that could be hunted. Fortunately, neither Malleefowl nor Brush-turkey figured among these latter species, and thus, for the first time, both were fully protected by law.

Despite being granted continuous protection by law since 1918, all three mound-builders continued to suffer from hunting. (Figure 8.1 shows a hunting party in the 1900s near Cobar, NSW, displaying at least three shot Malleefowl). Although some people certainly hunted megapodes for sport, it is likely that most were shot for food. This was especially evident during the depression period, and several Australian cookbooks from that time contain recipes detailing how to prepare them for the dinner table. Mound-builders are an easy target to hunt, first, because they can reliably be found near the incubation mounds, and second, because when fleeing from predators, they tend to fly conspicuously into trees. Since the late 1900s, however, hunting has become a minor threat compared to habitat destruction and predation by introduced predators, as outlined further below. For Brush-turkeys, the cessation of hunting pressure has been accompanied by the recovery of many populations, especially near urban centres.

Interestingly, Scrubfowl appear to have rarely been hunted, although their eggs are a major source of protein. This avoidance of hunting adult Scrubfowl is widespread throughout the distribution of the genus, also outside Australia. Although unconfirmed, there are anecdotal accounts that Scrubfowl putrefy extremely rapidly after death, which may have resulted in a general aversion to using these birds as food. We have been unable to determine whether similar attitudes exist among aborigines living with Scrubfowl in northern Australia.

## Conservation of the Malleefowl

The predicament of the Malleefowl represents a sad chapter in Australian history. Originally, this species occurred over large areas of southern, central and western Australia, and was widespread in every mainland state except Queensland. Today, it is restricted to patches of suitable habitat in West Australia, South Australia, Victoria, and New South Wales. Its status in the Northern Territory is generally assumed to be extinct, although recent unconfirmed sightings suggest it may still occur in the south-west region[4]. In New South Wales, the species was already disappearing from many parts by the year 1916, and in 1985, the population was estimated at roughly only 750 breeding pairs.[5] A 1999 survey in South Australia showed that less than half of all grid cells (size one degree) occupied by Malleefowl prior to 1981 still contained the species.[6] The Malleefowl population in Victoria was estimated at less than 1000 pairs in 1995, with the national population at no more than 2000 pairs. Legally, the species is now listed as *Endangered* in both Victoria and New South Wales, as *Vulnerable* in South Australia, as *Critically Endangered* in the Northern Territory and as *Fauna that is Rare Or Is Likely To Become Extinct* in Western Australia. The most detailed assessment of the status of the species compared data collected between 1980 and 2005 from 64 reliably studied sites throughout its range.[7] The trends varied considerably between States, with the decline being greatest in South Australia and less dramatic in Western Australia, with even a possible increase in New South Wales. However, overall, this site-based survey demonstrated unequivocal declines in the national populations, including reserves set aside for Malleefowl.

The factors contributing to the decline of the Malleefowl during the first half of the 20th century were accurately described by F. Lewis.[8] He mentioned clearing of the mallee as the main threat, including widespread burning of the forest by graziers who wanted to provide grass for their sheep. Some areas were also leased to eucalyptus oil distillers, who often cut down whole trees to obtain the oil. Lewis also mentioned that this species suffered from the removal of eggs by local farmers, the killing of chicks by foxes and the fact that significant numbers of birds were caught in rabbit traps.

Today, habitat destruction and predation by foxes are still the main threats for the species, while rabbit traps and egg consumption are of less concern. In addition, it is now recognised that Malleefowl also suffer from food shortage through competition with goats, rabbits and domestic stock.[9] Malleefowl habitat is destroyed or degraded in three ways: it is cleared, fragmented or modified for the cultivation of wheat and other crops; it is grazed by introduced animals, especially sheep, cattle, rabbits and goats; and finally, it is destroyed by frequent fires. Although fires are a natural and necessary phenomenon in most Australian landscapes, they are now occurring at much higher frequencies than prior to European history. Mallee is amongst the most flammable of vegetation types, and large fires consuming hundreds of thousands of hectares are now occurring at approximately 20-year cycles in most mallee in south-eastern Australia.[4] Joe Benshemesh showed that mallee habitat is most suitable for the Malleefowl about 40–60 years after fire, whereas the quality of this habitat is low or unsuitable for the species during the first 30 years after fire.[10] Common farming practices today, however, include fires at intervals of 20 years or less, to produce optimal food for domestic stock.

Red foxes take a heavy toll on Malleefowl, as they do on many ground-living creatures in Australia. They have the greatest impact on this species in small reserves or areas where the habitat is of marginal quality. In these locations, they often dig up large proportions of the eggs and kill many chicks and juveniles, and less often adults. In one study by David Priddel and Robert Wheeler, foxes killed half of a total of 100 chicks within four months of their release.[9] These workers estimated that by destroying eggs and killing chicks, foxes may reduce the reproductive output of a population by up to 75%. However, in larger areas with suitable habitat, foxes may not have as great an impact on the species as assumed. In such areas in Victoria, Joe Benshemesh found that breeding densities were higher when there was a good amount of vegetation cover above 2 m, while he could not find any correlation to the abundance of foxes or food-bearing shrubs.[11] This result indicates that sufficient vegetation cover is the main key to the survival of the species.

There is, however, considerable hope for the Malleefowl. State governments and local conservation organisations are engaged in several conservation projects that aim to bring the Malleefowl back from the brink of extinction. In all states, a small army of remarkably dedicated volunteers, coordinated by the National Malleefowl Recovery Team, carefully monitor a large number of sites using continually refined survey methods. These volunteers provide almost all the data essential for developing conservation plans. This work is especially evident in Western Australia, where numerous community based Malleefowl groups work with the private landholders on whose land most of the state's Malleefowl populations are found.

Conservation reserves containing Malleefowl have been established in four states. Those areas which are not yet part of the reserve system are increasingly covered by special conservation agreements with private landholders, even though this percentage still needs to be considerably increased. On Aboriginal Land, Indigenous Protected Areas provide recognition of conservation values to large areas that support Malleefowl. Indeed, recent surveys in the Anangu Pitjantjatjara lands of north-western South Australia, provide some positive news.[12] These regions, which probably contain the largest tracks of relatively undisturbed Malleefowl habitat in the entire country, appear to support excellent numbers of the species, though in naturally low densities. The impacts of too frequent fire on Malleefowl, however, remains a serious issue. Frequent burning may more than double the productivity of mallee habitats for sheep and other stock, while destroying Malleefowl habitat. This aspect of habitat management clearly needs to be addressed in any future management plans.

Joe Benshemesh has prepared a recovery plan for this species for two periods, 2000–2005[6] and 2006–2010,[4] both of which outline the necessary actions aimed at conserving the species in detail. Most actions are aimed at reversing the negative impacts of habitat fragmentation, which is now seen as one of the major causes of decline for this species. The most critical habitats for the survival of this species are in the semi-arid and arid zones: shrublands and low woodlands dominated by mallee, and associated habitats such as Broombush *Melaleuca uncinata*, Scrub Pine *Callitris verrucosa*, Red Ironbark *Eucalyptus sideroxylon* or Mulga *Acacia aneura*[4]. Let us hope that sufficiently large areas covered in these habitats will continue to exist, and thus support Malleefowl.

## Conservation of the Australian brush-turkey

Many people living along the East Coast of Australia encounter Brush-turkeys daily, as some of these birds have adapted well to living near human settlements. The more frequent sightings of 'suburban Brush-turkeys' may suggest that this species has become more common than it used to be. However, we would like to express this opinion with caution, as there are also indications that this species has disappeared from many areas it formerly occupied, and the perception of abundance is only seeing one part of the picture.

A recent review assessed 1560 reports of Brush-turkeys collected between 1788 and 2004, from journals, museum collections, and various databases and fauna atlases throughout Australia.[13] This study showed that the species formerly occurred as far south as Jindabyne in the Snowy Mountains and as far west as the Pilliga region. Today, it is no longer found in these regions and has withdrawn to a range that only extends to the Illawarra region (south of Sydney) to the south and the Great Dividing Range and Nandewar Range to the west.[13] The former wider distribution of the species may partly have been favoured by the introduction of the prickly pear early in the 20th century, a cactus that covered large parts of the country. Brush-turkeys most likely fed on the fruit of this plant and benefited from the cover it provided. When the cactus was decimated, Brush-turkeys also seem to have withdrawn their range. However, some historic records in the far south and west, from places such as Narrabi, in 1880s, Nyngan in 1900, and Cape Howe in the 1800s, greatly pre-date the introduction of the Prickly Pear.[13] This indicates that the species's former distribution cannot be explained by the introduction of the cactus alone.

One smaller Brush-turkey population in the Nandewar and Brigalow Belt has now even become so isolated from the remaining population that it has been listed as an endangered population under the NSW *Threatened Species Act* 1995. The main reason for listing this population is its vulnerability to habitat loss and modification caused by alterations of natural fire regimes, clearing and grazing by herbivores. Additionally, the population suffers from predation by cats, foxes and wild dogs, as well as wild pigs digging through its incubation mounds. While these threats are only officially listed as threatening processes for this isolated population, they do also threaten Brush-turkey populations in many other areas of Australia and have most likely caused the decline of population numbers since Europeans first occupied Australia.

Even suburban Brush-turkeys today usually only move into suburban areas that are adjacent to forested areas, indicating that this species still depends on the availability of natural habitat. In Brisbane, for example, the majority of the urban Brush-turkey population occur in suburbs adjacent to large areas of bushland which contains dry and wet rainforests and gullies, and it is only during recent years that some individuals have moved into suburbs more distant from bushland refuges.[14] Similarly, in the northern suburbs of Sydney and along the NSW Central Coast, these birds are predominantly found in areas close to natural bushland. A recent radio-tracking study in these regions showed that suburban Brush-turkeys usually spend only a minority of their time in the suburbs and mainly remain in the surrounding natural bushland in between.[15] These examples show that even though Brush-turkeys are now more frequently seen in suburban areas, they nevertheless depend on the availability of natural bushland for their survival.

## Conservation of the Orange-footed Scrubfowl

The Scrubfowl is the only species for which we have been able to gain insights into its range before European colonisation of Australia. This is based on investigations of the apparently ancient 'hillocks' scattered widely throughout the savannas of northern Australia, often at considerable distance from the current more coastal habitats of the species. Although some scientists have interpreted these relatively large structures to be of human origin,[16] there is considerable evidence to suggest that they are actually very old Scrubfowl incubation mounds.[17] It is now believed that Scrubfowl formerly occurred much further inland in areas that used to be covered with forests, but due to long-term climate change and associated changes in vegetation cover, they withdrew to coastal areas where they are found today.[18]

Of the three Australian species, the Scrubfowl seems to have been the least impacted by European influences. Under current Commonwealth legislation, the conservation status of the Scrubfowl is listed as 'least concern'. This is largely due to the fact that Scrubfowl live predominantly in tall rainforests and vine thickets, often in remote areas along the northern coasts of Australia where larger areas of this type of habitat still exist. However, in areas where human populations are greater, such as in Queensland near Cooktown and Port Douglas and the adjacent highlands, Scrubfowl suffer increasingly from habitat destruction. In these areas, clearing for settlement, grazing and agriculture has removed much of the suitable habitat, especially from coastal regions and the Atherton

Tablelands, where the species is now confined to small forest fragments. As a result, the Royal Australasian Ornithologists Union listed this eastern Queensland population as 'of special concern' in 1992.

In recent years, Scrubfowl have increasingly been sighted in the suburbs of urban settlements, including the cities of Darwin, Cairns and Port Douglas. Here, they usually occupy well vegetated gardens in the more established suburbs. As with Brush-turkeys, these increased sightings do not necessarily reflect an increase in population numbers, as the species nevertheless depends on the availability of sufficiently large patches of native vegetation for survival. Furthermore, Scrubfowl now often share their environment with feral pigs, which are becoming increasingly common in Northern Australia. Pigs are known to destroy incubation mounds while digging through them for eggs, and are a serious threat for Scrubfowl in many areas.

## Living with suburban mound-builders

Sightings of Brush-turkeys – and to some extent Scrubfowl – in suburban areas have increased during the last decade or so. Mostly, these birds visit suburban gardens while searching for food, but some individuals come to stay and build their incubation mound. While some people see this as an opportunity to watch a fascinating species up close, others regard these large birds as disruptive and untidy invaders of their intimate home territory.[19] Most complaints from suburban residents focus on the destruction of their gardens by the mound-builders and on the mess these birds make while building their large mounds and searching for food. A mound comprised of several tons of organic matter may, after all, quickly include all of the available mulch, lawn clippings, newly planted seedlings, pot plants and even sprinkler systems! Hence, a garden in which a mound-builder has decided to settle typically looks rather different after a few days of mound-building. In addition, people complain about Brush-turkeys making noise while travelling over corrugated iron roofs and of stealing pet food. Furthermore, where Brush-turkeys congregate in high numbers, they can strip the topsoil layer and prevent the regrowth of young plants.

Several factors explain why these birds are attracted to our suburbs. First, since the 1970s, gardens have included increasing numbers of native plant species, often creating rainforest-like conditions. They thus provide the semi-shade that mound-builders prefer when building their mounds. Also, by watering their gardens and covering their plants with mulch, people create a micro-environment that provides an abundance of invertebrate food for mound-builders as well as building material for their mounds. While Australia is in the grip of an increasingly severe drought, these moist garden environments appear to be important refuges for mound-building birds; without moist material, an incubation mound cannot function properly.

Second, it appears that since hunting has mostly ceased, mound-builders have become less wary of humans and more willing to approach. Even as recently as the 1960s, Brush-turkeys tended to be described as being excessively shy throughout most of their range. While this is still true in undisturbed environments, most Brush-turkeys in urban areas are now fully habituated to the presence of people. Third, some wildlife managers believe that the increased fox baiting near suburban settlements has helped mound-builders to survive in such areas. However, this has not been fully assessed and may be balanced by the increased mortality rate caused by feral cats. Fourth, suburbs of many cities continue to expand into areas inhabited by mound-builders. In this case, it is humans rather than the birds who should be called the invaders.

Finally, an additional factor became evident during a recent study on the Central Coast near Sydney.[15] In this investigation, people responded to a questionnaire that explored their attitudes towards Brush-turkeys. While few people admitted to feeding Brush-turkeys directly, half of the 140 respondents provided food either via an uncovered compost heap, a bird feeder from which seed dropped to the ground, or via pet or chicken food that was left outside. The Brush-turkeys radio-tracked in this study frequented such backyards for feeding on a daily basis, although they spend most of their day in nearby bushland. This level of utilisation of human-provided foods is likely to be typical of many other suburban areas.

What can be done to mitigate the disturbance caused by suburban mound-builders? Translocating the birds to other areas is often not a feasible, ecologically sensible or ethical solution. Recent translocations of radio-tracked Brush-turkeys in northern Sydney showed that these immediately attempted to return, and two adult males involved were killed soon after release, one by a car. Such translocation causes unjustifiable stress on the birds and may lead to their death. Destruction of the incubation mounds is only legal while these do not contain any eggs or chicks, and this is often difficult to assess. Furthermore, even when mounds are removed, a Brush-turkey male determined to construct his mound at a certain location almost always rebuilds quickly. Hence, many people living with these birds' mounds in their backyard eventually are obliged to accepting Brush-turkeys as their guests.

Some measures can help to protect the gardens visited by mound-builders. First, if living in an area where these birds occur, plan your new garden with mound-builders in mind. New gardens are particularly attractive – and most vulnerable – to these birds. By using a heavy covering such as river gravel over standard mulch, the area becomes much less attractive for the birds. Tree guards are useful to prevent raking of newly planted trees or other plants, or you may even consider fencing off new plantings or vegetable gardens altogether. Also, develop your garden in stages, which makes it more resistant to serious damage if you protect plants at each stage. And, most importantly, do not attract the birds with bird feeders, pet food, or uncovered compost heaps.

We also hope that with this book, we can increase the awareness about how special mound-builders are, and that being able to watch them in your own backyard can be fascinating entertainment. Such awareness and entertainment may help to endure the negative aspects described above. Some studies have indeed shown that where people received more information about these birds, they developed a more positive attitude towards them.[15,19] For example, a mound that was seen as an 'unsightly mess' had become a 'nest being cared for by an attentive parent'. We also need to keep in mind that while living in suburbia, these birds face many apparent threats, such as roads and traffic, a lack of protective thickets, and a high density of cats.[14] Predation of cats often removes almost all chicks, and survival rates are only above zero in those areas where chicks can hide in dense thickets.[20] By providing suitable incubation and hiding conditions in an otherwise drought stricken area, we may thus well mitigate such negative impacts to at least some extent.

## A final comment

We started this book by referring to the legacy of Harry Frith's foundational studies in tough mallee country near Griffith in central New South Wales. After summarising almost a decade of extraordinary research in his book *The Mallee-Fowl,* Frith concluded with a plea for some of the rapidly disappearing mallee of the area to be preserved to allow the birds to remain. Soon after, Pulletop Nature Reserve was declared, still containing a few malleefowl pairs. Today, this mallee remnant is a tiny island in a sea of wheat fields. And although they held on for decades, the last birds residing there have finally died. Frith's legacy of study and determination, however, has not ceased and continues to inspire new generations to work toward understanding these astonishing birds so that they may be preserved: science for conservation. To conclude in Harry's own words:

> *Surely a little effort now is worthwhile to ensure that the malleefowl goes on with its work in the inland scrubs, to the joy of future generations.* (H. Frith 1962: 130).[21]

# ENDNOTES

## Chapter 1: Familiar yet distinct


1 Frith HJ (1962) *The Mallee-Fowl: The Bird that Builds an Incubator.* Angus & Robertson, Sydney.
2 Pigafetta A (1521) In *Primo Viaggio intorno al Globo.* (Ed. S Amoretti). Milano.
3 Buffon GLL (1770–1786) *Histoire Naturelle des Oiseaux.* De L'Imprimerie Royale, Paris.
4 Dekker RWRJ and Wattel J (1987) Egg and image: new and traditional uses for the maleo (*Macrocephalon maleo*). In *The Value of Birds.* (Eds AW Diamond and FL Filion) pp. 83–87. ICBP Technical Publication 6. BirdLife International, Cambridge.
5 Steadman DW (2006) *Extinction and Biogeography of Tropical Pacific Birds.* University of Chicago Press, Chicago.
6 Latham J (1824) *General History of Birds.* Vol. 10. Leigh & Sotheby, London.
7 Gould J (1848) *Birds of Australia.* Vol. 1. Taylor, London.
8 Dumont C (1840) Megapode. In *Dictionaire des Sciences Naturelles* **29,** 414–418. Levrault, Paris.


## Chapter 2: Taxonomy, distribution and habitat


1 Frith HJ (1959) Incubator birds. *Scientific American* **201**, 52–58.
2 Clark GA (1964) Life histories and the evolution of megapodes. *Living Bird* **3**, 149–167.
3 Jones DN, Dekker RWRJ and Roselaar CS (1995) *The Megapodes.* Oxford University Press, Oxford.
4 Birks SM and Edwards SV (2002) A phylogeny of the megapodes (Aves: Megapodiidae) based on nuclear and mitochrondrial DNA sequences. *Molecular Phylogenies and Evolution* **23**, 408–421.
5 Brom TG and Dekker RWRJ (1992) Current studies on megapode phylogeny. In *Proceeding of the First International Megapode Symposium.* (Eds RWRJ Dekker and DN Jones). pp. 7–17. Christchurch, New Zealand, December 1990. Zoologische Verhandelingen, Amsterdam.

6 Rich PV and van Tets GF (1985) *Kadimakara: Extinct Vertebrates of Australia*. Pioneer Design Studios, Lilydale, VIC.
7 Steadman DW (2006) *Extinction and Biogeography of Tropical Pacific Birds*. University of Chicago Press, Chicago.
8 Boles WE and Ivison TJ (1999) A new genus of dwarf megapode (Galliformes: Megapodiidae) from the late Oligocene of Central Australia. *Smithsonian Contributions to Paleobiology* **89**, 199–206.
9 Kloska C and Nicolai J (1988) Fortpflanzungsverhalten des Kamm-Talegalla (*Aepypodius arfakianus*). *Journal für Ornithologie* **129**, 185–204.
10 Mauro I (2005) Field discovery, mound characteristics, bare parts, vocalisations and behaviour of Bruijn's Brush-turkey (*Aepypodius bruijnii*). *Emu* **105**, 273–281.
11 Brookes GB (1919) Report on investigations in regard to the spread of prickly-pear by the Scrub Turkey. *Emu* **18**, 288–292.
12 Göth A, Nicol KP, Ross G and Shields J (2006) Present and past distribution of Australian brush-turkeys *Alectura lathami* in New South Wales – implications for management. *Pacific Conservation Biology* **12**, 22–30.
13 Howard T and Crawford I (1999) The Australian brush-turkey in the ACT: Further discussion. *Canberra Bird Notes* **24**, 173–176.

## Chapter 3: Appearance and ecology

1 Barrett C (1931) Avian mound builders and their mounds. *Bulletin of the New York Zoological Society* **34**, 107–129.
2 Olsen V (2002) Evolution of Avian Caratenoid pigmentation: behavioural, biochemical and comparative approaches. PhD thesis, University of Queensland, Brisbane.
3 Jones DN (1987). Behavioural ecology of reproduction in the Australian Brush-turkey *Alectura lathami*. PhD thesis, Griffith University, Brisbane.
4 Frith HJ (1959b) Breeding of the Mallee Fowl, *Leipoa ocellata* Gould (Megapodiidae). *CSIRO Wildlife Research* **4**, 31–60.
5 Böhner J and Immelmann K (1987) Aufbau, Variabilität und mögliche Funktion des Rufduetts beim Thermometerhuhn *Leipoa ocellata*. *Journal für Ornithologie* **128**, 91–100.
6 Crome FHJ and Brown HE (1979) Notes on social organisation and breeding of the Orange-footed Scrubfowl *Megapodius reinwardt*. *Emu* **79**, 111–119.

7 Göth A, Vogel U and Curio E (1999). The acoustic communication of the Polynesian megapode *Megapodius pritchardii* G. R. Gray. *Zoologische Verhandelingen* **327**, 37–51.

8 Frith HJ (1962) *The Mallee-Fowl: The Bird that Builds an Incubator.* Angus & Robertson, Sydney.

9 Benshemesh J (1992) The conservation ecology of the Malleefowl, with particular regard to fire. PhD thesis, Monash University, Clayton.

10 Harlen R and Priddel D (1996) Potential food resources available to malleefowl *Leipoa ocellata* in marginal mallee lands during drought. *Australian Journal of Ecology* **21**, 418–428.

11 Reichelt R and Jones DN (2008) Long-term observations of the diet of the malleefowl *Leipoa ocellata* near the Little Desert, Western Victoria. *Australian Field Ornithology* **25**, 22–30.

12 Healey C (1994) Dispersal of newly hatched Orange-footed Scrubfowl *Megapodius reinwardt. Emu* **94**, 221.

## Chapter 4: The mound

1 Barrett C (1931) Avian mound builders and their mounds. *Bulletin of the New York Zoological Society* **34**, 107–129.

2 Russell A (1944) *Bush Ways.* Australian Publications, Sydney.

3 Weathers WW, Seymour RS and Baudinette RV (1993) Energetics of mound-tending behaviour in the malleefowl, *Leipoa ocellata* and brush turkey *Alectura lathami* (Megapodiidae). *Animal Behaviour* **94**, 134–150.

4 Jones DN (1988) Construction and maintenance of the incubation mounds of the Australian Brush-turkey *Alectura lathami. Emu* **88**, 210–218.

5 Seymour RS (1995) Calorimetic investigation on mound-building birds. *Acta Thermochimica* **250**, 319–328.

6 Palmer C, Christian KA and Fisher A (2000) Mound characteristics and behaviour of the Orange-footed Scrubfowl in the seasonal tropics of Australia. *Emu* **100**, 54–63.

7 Stone T (1991) Megapode mounds and archaeology in northern Australia. *Emu* **91**, 255–256.

8 Crome FHJ and Brown HE (1979) Notes on social organisation and breeding of the Orange Footed Scrubfowl *Megapodius reinwardt. Emu* **79**, 111–119.

9 Vleck D, Vleck CM and Seymour RS (1984) Energetics of embryonic development in the megapode birds, mallee fowl *Leipoa ocellata* and brush turkey *Alectura lathami*. *Physiological Zoology* **57**, 444–456.
10 Booth DT (1987) Effect of temperature on development of Mallee Fowl *Leipoa ocellata* eggs. *Physiological Zoology* **60**, 437–445.
11 Jones DN (1987) Behavioral ecology of reproduction in the Australian Brush-turkey *Alectura lathami*. PhD. thesis, Griffith University, Brisbane.
12 Göth A (2007) Incubation temperatures and sex ratios in Australian brush-turkey (*Alectura lathami*) mounds. *Austral Ecology* **32**, 378–385.
13 Seymour RS and Bradford DF (1992) Temperature regulations in the incubation mounds of the Australian Brush turkey. *Condor* **94**, 134–150.
14 Seymour RS, Vleck D and Vleck CM (1986) Gas exchange in the incubation mounds of megapode birds. *Journal of Comparative Physiology B*. **156**, 773–782.
15 Jones DN (1990) Male mating tactics in a promiscuous megapode: patterns of incubation mound ownership. *Behavioral Ecology* **1**, 107–115.
16 Göth A and Astheimer L (2006) Development of mound building in Australian Brush-turkeys (*Alectura lathami*): the role of learning, testosterone and body mass. *Australian Journal of Zoology* **54**, 71–78.
17 Jones DN (1987) Selection of incubation mound sites by the Australian Brush-turkey *Alectura lathami*. *Ibis* **130**, 251–260.
18 Bowman DMJS, Woinarski JCZ and Russell-Smith J (1994) Environmental relationships of orange-footed scrubfowl nests in the Northern Territory. *Emu* **94**, 181–185.
19 Priddel D and Wheeler R (2005) Fecundity, egg size and the influence of rainfall in an isolated population of malleefowl (*Leipoa ocellata*). *Wildlife Research* **32**, 639–648.
20 Hercock M (2004) The impacts of recreation and tourism in the remote North Kimberley region of Western Australia. *The Environmentalist* **19**, 259–275.

## Chapter 5: Abandoned eggs

1 Chisholm AH (1934) *Bird Wonders of Australia*. Angus and Robertson, Sydney.
2 Frith HJ (1962) *The Mallee-Fowl: The Bird that Builds an Incubator*. Angus & Robertson, Sydney.

3 Booth DT (1988) Shell thickness in megapode eggs. *Megapode Newsletter* **2**, 13.
4 Seymour RS, Vleck D, Vleck CM and Booth DT (1987) Water relations of buried eggs of mound building birds. *Journal of Comparative Physiology B.* **157**, 413–422.
5 Dekker RWRJ and Brom TG (1990) Maleo eggs and the amount of yolk in relation to different incubation strategies in Megapodes. *Australian Journal of Zoology* **38**, 19–24.
6 Jones DN, Dekker RWRJ and Roselaar CS (1995) *The Megapodes.* Oxford University Press, Oxford.
7 Vleck D, Vleck CM and Seymour RS (1984) Energetics of embryonic development in the megapode birds, mallee fowl *Leipoa ocellata* and brush turkey *Alectura lathami. Physiological Zoology* **57**, 444–456.
8 Göth A (2007) Mound and mate choice in a polyandrous megapode: females lay more and larger eggs in nesting mounds with the best incubation temperatures. *The Auk* **124**, 253–263.
9 Christians JK (2000) Trade-offs between egg size and number in waterfowl: An interspecific test of the van Noordwijk and de Jong model. *Functional Ecology* **14**, 497–501.
10 Priddel D and Wheeler R (2005) Fecundity, egg size and the influence of rainfall in an isolated population of malleefowl (*Leipoa ocellata*). *Wildlife Research* **32**, 639–648.
11 Frith HJ (1959) Breeding of the Mallee Fowl, *Leipoa ocellata* Gould (Megapodiidae). *CSIRO Wildlife Research* **4**, 31–60.
12 Jones DN (1988) Hatching success of the Australian Brush-turkey *Alectura lathami* in South-East Queensland. *Emu* **88**, 260–263.
13 Fleay DH (1937) Nesting habits of the Brush Turkey. *Emu* **36**, 153–163.
14 Coles C (1937) Some observations on the habits of the Brush Turkey (*Alectura lathami*). *Proceedings of the Zoological Society London* **107A**, 261–273.
15 Brickhill J (1987) Breeding success of Malleefowl *Leipoa ocellata* in Central New South Wales. *Emu* **87**, 42–45.
16 Benshemesh J (1992) The conservation ecology of Malleefowl, with particular regard to fire. PhD Thesis, Monash University, Clayton.
17 Booth DT (1987) Home range and hatching success of Mallefowl *Leipoa ocellata* Gould (Megapodiidae) in Murray Mallee near Renmark, S.A. *Wildlife Research* **14**, 95–104.

18 Crome FHJ and Brown HE (1979) Notes on social organisation and breeding of the Orange Footed Scrubfowl *Megapodius reinwardt. Emu* **79**, 111–119.
19 Booth DT and Jones DN (2002). Underground nesting in megapodes. In *Avian Incubation: Behaviour, Environment, and Evolution.* (Ed. DC Deeming), pp. 192–206. Oxford University Press, Oxford.
20 Göth A and Booth DT (2005) Temperature-dependent sex-ratio in a bird. *Biology Letters* **1**, 31–33.
21 Göth A (2007) Incubation temperatures and sex ratios in Australian brush-turkey (*Alectura lathami*) mounds. *Austral Ecology* **32**, 378–385.
22 Seymour RS, Vleck D and Vleck CM (1986) Gas exchange in the incubation mounds of megapode birds. *Journal of Comparative Physiology B.* **156**, 773–782.

## Chapter 6: Growing up without parental care

1 Gould J (1842) On the habits of the *Alectura lathami. Tasmanian Journal of Natural Science* **1**, 21–24.
2 Russell A (1944) *Bush Ways.* Australian Publications, Sydney.
3 Seymour RS and Ackerman RA (1980) Adaptations to underground nesting in birds and reptiles. *American Zoologist* **20**, 437–444.
4 Göth A (2002) Behaviour of Australian Brush-turkey (*Alectura lathami,* Galliformes: Megapodiidae) chicks following underground hatching. *Journal für Ornithologie* **143**, 477–488.
5 Booth DT (1985) Thermoregulation in neonate Brush Turkeys (*Alectura lathami*). *Physiological Zoology* **58**, 374–379.
6 Benshemesh J (1992). The conservation ecology of Malleefowl, with particular regard to fire. PhD Thesis, Monash University, Clayton.
7 Göth A and Vogel U (2003) Juvenile dispersal and habitat selectivity in the megapode *Alectura lathami* (Australian Brush-turkey). *Wildlife Research* **30**, 69–74.
8 Healey C (1994) Dispersal of newly hatched Orange-footed Scrubfowl *Megapodius reinwardt. Emu* **94**, 221.
9 Göth A and Vogel U (2002) Chick survival in the megapode *Alectura lathami* (Australian Brush-turkey). *Wildlife Research* **29**, 503–511.
10 Göth A and Jones DN (2001). Transmitter attachment and its effects on Australian brush-turkey hatchlings. *Wildlife Research* **28**, 73–78.
11 Priddel D and Wheeler R (1994) Mortality of captive-raised Malleefowl, *Leipoa ocellata,* released into a mallee remnant within the wheatbelt of New South Wales. *Wildlife Research* **21**, 543–552.

12 Göth A and Proctor H (2002) Pecking preferences in hatchlings of the Australian Brush-turkey: the role of food type and colour. *Australian Journal of Zoology* **50**, 93–102.
13 Göth A (2001) Innate predator recognition in Australian Brush-turkey hatchlings. *Behaviour* **138**, 117–136.
14 Göth A (2001) Survival, habitat selectivity and behavioural development of the Australian brush-turkey *Alectura lathami*. PhD thesis, Griffith University, Brisbane.
15 Göth A and Jones DN (2003) Ontogeny of social behaviour in the megapode *Alectura lathami* (Australian Brush-turkey). *Journal of Comparative Psychology* **117**, 36–43.
16 Wong S (1999) Development of behaviour of hatchlings of the Australian brush-turkey *Alectura lathami*. PhD Thesis, Griffith University, Brisbane.
17 Göth A and Evans CS (2004) Social responses without early experience: Australian brush-turkey chicks use specific visual cues to aggregate with conspecifics. *Journal for Experimental Biology* **207**, 2199–2208.
18 Barry K and Göth A (2006) Call recognition in chicks of the Australian brush-turkey (*Alectura lathami*). *Animal Cognition* **9**, 47–54.
19 Booth DT (1989) Growth rates of Malleefowl and an Australian Brush-turkey in captivity. *Megapode Newsletter* **3**, 4.
20 Göth A and Astheimer L (2006) Development of mound building in Australian Brush-turkeys (*Alectura lathami*): the role of learning, testosterone and body mass. *Australian Journal of Zoology* **54**, 71–78.
21 Jones DN (1990) Male mating tactics in a promiscuous megapode: patterns of incubation mound ownership. *Behavioral Ecology* **1**, 107–115.
22 Priddel D and Wheeler R (2005) Fecundity, egg size and the influence of rainfall in an isolated population of malleefowl (*Leipoa ocellata*). *Wildlife Research* **32**, 639–648.
23 Frith HJ (1959) Breeding of the mallee fowl, *Leipoa ocellata* Gould (Megapodiidae). *CSIRO Wildlife Research* **4**, 31–60.

## Chapter 7: Social and reproductive behaviour

1 Frith HJ (1962) *The Mallee-Fowl: The Bird that Builds an Incubator*. Angus & Robertson, Sydney.
2 Bennett KH (1884) On the habits of the mallee hen, *Leipoa ocellata. Proceeding of the Linnean Society of NSW* **8**, 193–197.
3 Jones DN (1979) Notes of the breeding habits of the Brush Turkey. *Sunbird* **10**, 8–10.

4 Jones DN (1988) Selection of incubation mound sites by the Australian Brush-turkey. *Ibis* **130**, 251–260.
5 Jones DN (1990) Male mating tactics in a promiscuous megapode: patterns of incubation mound ownership. *Behavioural Ecology* **1**, 107–115.
6 Jones DN (1988) Construction and maintenance of the incubation mounds of the Australian Brush-turkey. *Emu* **88**, 210–218.
7 Jones DN (1994) Reproduction without parenthood: Male tactics and female choice in a promiscuous bird. In *Animal Societies: Individuals, Interactions and Organisation.* (Eds PJ Jarman and A Rossiter). pp. 135–146. Kyoto University Press, Kyoto.
8 Jones DN (1990) Social organization and sexual interactions in Australian Brush-turkeys (*Alectura lathami*): implications of promiscuity in a mound-building megapode. *Ethology* **84**, 89–104.
9 Birks S (1997) Paternity in the Australian Brush-turkey, *Alectura lathami*, a megapode bird with uniparental male care. *Behavioural Ecology* **8**, 560–568.
10 Birks S (1996) Reproductive behaviour and paternity in the Australian Brush-turkey, *Alectura lathami*. PhD thesis, Cornell University, Ithaca, USA.
11 Göth A (2007) Mound and mate choice in a polyandrous megapode: females lay more and larger eggs in nesting mounds with the best incubation temperatures. *Auk* **124**, 253–263.
12 Baltin S (1969) Zur Biologie und Ethologie des Talegalla-Huhnes (*Alectura lathami* Grey) unter besonderer Berücksichtigung des Verhaltens während der Brutperiode. *Zeitschrift für Tierpsychologie* **26**, 524–572.
13 Dow DD (1988) Sexual interactions by Australian Brush-turkeys away from the incubation mound. *Emu* **49**, 49–50.
14 Weather WW and Seymour RS (1998) Behaviour and time-activity budgets of Malleefowl *Leipoa ocellata* in South Australia. *Emu* **98**: 288–296.
15 Frith HJ (1955) Incubation in the mallee fowl (*Leipoa ocellata*, Megapodiidae). In *Acta XI Congressus Internationalis Ornithologici* Basel, 1954. (Eds A Portmann and E Sutter). pp. 570–574. Birkhauser, Basel.
16 Böhner J and Immelmann K (1987) Aufbau, Variabilität und mögliche Funktionen des Rutduetts beim Thermometerhuhn *Leipoa ocellata. Journal für Ornithologie* **128**, 91–100.
17 Immelmann K and Böhner J (1984) Beobachtungen am Thermometerhuhn (*Leipoa ocellata*) in Australien. *Journal für Ornithologie* **125**, 141–155.

18 Weathers WW, Weathers DL and Seymour RS (1990). Polygyny and reproductive effort in the Malleefowl, *Leipoa ocellata*. *Emu* **90**, 1–6.
19 Göth A and Vogel U (2004) Is monogamy in the Polynesian megapode (*Megapodius pritchardii*) related to its high relative egg weight? *Auk* **121**, 308–317.
20 Palmer C, Christian KA and Fisher A (2000) Mound characteristics and behaviour of the orange-footed scrubfowl in the seasonal tropics of Australia. *Emu* **100**, 54–63.
21 Crome FHJ and Brown HE (1979) Notes on social organization and breeding of the orange-footed scrubfowl *Megapodius reinwardt*. *Emu* **79**, 111–119.

## Chapter 8: Conservation and management of Australian mound-builders

1 Chisholm AH (1943) *Bird Wonders of Australia*. Angus and Robertson, Sydney.
2 Barrett C (1932). Avian mound builders and their mounds. *Bulletin New York Zoological Society* **34**, 107–129.
3 Jarman PJ and Brock MA (2004) Is it just an act? Threatened species legislation in NSW. In *Threatened Species Legislation: Is it just an Act?* (Eds P Hutchings, D Lunney and C Dickman). pp 1–19. Royal Zoological Society of New South Wales, Mosman, NSW.
4 Benshemesh J (2007) Draft National Recovery Plan for Malleefowl 2006–2010. Department of Environment and Heritage, South Australia, Adelaide.
5 Barrett C (1943) *An Australian Animal Book*. Oxford University Press, Melbourne.
6 Benshemesh J (2000) National Recovery Plan for Malleefowl. Department of Environment and Heritage, South Australia, Adelaide.
7 Benshemesh J, Barker R and MacFarlane R (2007). Trend analysis of Malleefowl Monitoring Data. Report to Mallee CMA, Victorian Malleefowl Recovery Group and National Malleefowl Action Project, Melbourne.
8 Lewis F (1950) Some factors relating to the survival or otherwise of the Lowan or Mallee Fowl. *Victorian Naturalist* **67**, 142–143.
9 Priddel D and Wheeler RJ (1990) Conservation of the endangered Malleefowl *Leipoa ocellata*. *Congressus Internationalis Ornithologici* (Supplement) **483**, 121–124.

10 Benshemesh J (1990) In *The Mallee Lands, a Conservation Perspective.* (Eds JC Noble, PJ Joss and GK Jones). pp. 206–211. CSIRO PUBLISHING, Melbourne.
11 Benshemesh J (1992) The conservation ecology of Malleefowl, with particular regard to fire. PhD. thesis, Monash University, Clayton.
12 Partridge T and Ward M (2007) Nganamara in South Australia's Aboriginal Lands. In *Proceedings of the Malleefowl Forum 2007,* Katanning, WA, 7–10 September 2007. (Ed. SJJF Davies) pp. 23–27. Edith Cowen University, Perth.
13 Göth A, Nicol KP, Ross G and Shields J (2006) Present and past distribution of Australian brush-turkeys *Alectura lathami* in New South Wales - implications for management. *Pacific Conservation Biology* **11**, 22–30.
14 Jones DN, Sonnenburg R and Sinden KE (2004) Presence and distribution of Australian Brush-turkeys in the Greater Brisbane Region. *Sunbird* **34**, 1–9.
15 Göth A (2007) Australian brush-turkeys, corridors and human-wildlife conflicts. Unpublished report to the Gosford City Council, Gosford.
16 Bailey GN (1977) Shell mounds, shell middens and raised beaches in the Cape York Peninsula. *Mankind* **11**, 132–143.
17 Stone T (1991) Megapode mounds and archaeology in northern Australia. *Emu* **91**, 255–256.
18 Bowman DMJS, Panton WJ and Head J (1999) Abandoned Orange-footed Scrubfowl (*Megapodius reinwardt*) nests and coastal rainforest boundary dynamics during the late Holocene in monsoonal Australia. *Quaternary International* **59**, 27–38.
19 Jones DN and Everding SE (1991) Australian Brush-turkeys in a suburban environment: implications for conflict and conservation. *Wildlife Research* **18**, 285–297.
20 Göth A and Vogel U (2002) Chick survival in the megapode *Alectura lathami* (Australian Brush-turkey). *Wildlife Research* **29**, 503–511.
21 Frith HJ (1962) *The Mallee-Fowl: The Bird that Builds an Incubator.* Angus & Robertson, Sydney.

# INDEX


age of first breeding 43
air bubble in eggs *see* eggs, air bubble
*Alectura lathami purpureicollis* 18–9
appearance, 18, 23, 25, 68
Australian Brush-turkey 18
  appearance 18
  attracting females to mounds 85–7
  booming 21–3
  Cape York subspecies 18–9
  conservation 94, 99–101
  distribution 13–14, 99–101
  mate choice 86–7
  mating system 84–7
  mound construction 35–6
  social behaviour, adult 81–91
  social behaviour, chick 67–76
  suburban 34, 100–1, 102–4
  tarsus length 20
  vocalisations 21–3
  wattle 18–9, 21–3
Australian Giant Megapode *see* *Progura gallinacea*

Benshemesh, J. 29, 74, 98, 99
Birks, S. 84, 86
booming (vocalisation), 21–5
Booth, D. 58, 78–9
breeding seasons 46–7
Brush-turkeys, non-Australian species 13–4
burrow-nesting megapodes 10

calls *see* booming; duetting; vocalisations
Canendo, W. 39, 48, 64
cats, and mound-builders 74, 76, 100, 103–4
chick
  appearance 68
  behaviour 70–8
  dispersal 73
  growth 78–9
  movement 73–4
  predation 74–6, 102
  special features 69
  species recognition 76–8
  survival 74 *see also* cats; foxes
  thermoregulation 69, 72
cloaca 21, *see also* sexing; phallus
clutch size 3, 59, 61–3, 86, *see also* laying interval
compost heaps as mounds 34
conservation, 94, 97–102
conservation status *see* legal status

distribution 13–16, 99–101
  *see also* megapode, distribution
drinking 29–30
duetting 24–7
Dwyer, P. 39

eggs
  adaptations of *see* underground nesting, adaptations to
  air bubble 59, 61
  appearance 58
  harvesting 94–5
  number per female 61–3
  predation 63
  shell thickness 58, 59

size 59, 60–1, 90
yolk content 59–60

flying 30–1, 68, 69, 72–3, 96, *see also* movements
food 27–30
foraging *see* food
foxes and mound-builders 63, 74, 98, 100, 103
Frith, H.J. 4, 5, 29, 45, 48, 79, 58, 104–5

Galliformes 8
Göth, A. 40, 74, 76, 84, 87
growth 78–9

habitat destruction 14, 95–6, 98, 101
hatching
parental care 83–4
process 68–72 *see also* eggs, air bubble
success 63
heat production in mounds 39–43 *see also* incubation temperatures
historic discovery 1–3
hunting 95–7

incubation period 59, 64–5
incubation temperatures 39–43, 48, 59, 64–5, 87

Jones, D. 43, 78, 84, 86

laying interval 61–3 *see also* clutch size
legal status 95
life-span 31

Malleefowl
appearance 23
booming 24–5
conservation 94, 97–9
distribution 14–15, 99–101
duetting 24–5
and fire 98
habitat management 99
mating system 37, 81, 87–9
mound construction 36–7
recovery plan 99
tarsus length 24
vocalisations 22, 23–5
mating systems 81–3, 84–91
megapode
definition 5
distribution 8, 9, 12–16, *see also* distribution
eggs 58–61
extinct species 12
genera 10, 11
and reptiles 7
species 10, 11
taxonomy 8, 9
monogamy 82 *see also* mating systems
moulting 31 *see also* staffelmauser
mounds,
ancient 35
composition 33–43
construction, 35–9, 46
comparisons 3
description 33–8
ownership and mating system 35–6
re-use over many years 45
sharing between species 38–9
site locations 41, 43–6
size 34
temperatures 39–43 *see also* incubation temperatures
usurpation 35

and weather prediction 47–8
movements 30–31

New Holland Vulture 2
*Ngawupodius minya* 12

*Opuntia* spp. *see* Prickly Pear
Orange-footed Scrubfowl
appearance 25
conservation 101–2
distribution 15–16, 101
duetting 26–7
mating system 89–91
mound construction 37–8
suburban 102–4
tarsus length 26
vocalisations 25–7

pair bonds *see* mating system
parental care 3, 45, 67, 83–4
Parsons, B. 45
phallus 20 *see also* cloaca; sexing
phylogeny 8–10
Pigafetta, A. 1
polygamy 82 *see also* mating systems
precocial 3, 59, 68–9
predation 20, 21, 30, 63, 73, 74–6, 96, 98, 100, 104
predator recognition 74–6
Prickly Pear 14, 28–9, 100
Priddel, D. 29, 46, 74, 70, 98
primitive features 7–8
*Progura gallinacea* 12

recovery plan, *see* Malleefowl, recovery plan

Reichelt, 'Whimpey' 29, 31
roosting in trees 30, 90

sex ratios 64–5
sexing 19–20, 23 *see also* cloaca; phallus
Seymour, R. 42, 70
size in extinct megapodes 12, 19–20
size *see* tarsus length; growth
social behaviour
adult 81–91
chick 67–76
species recognition 76–8
staffelmauser 30–1
suburban interaction 34, 100–1, 102–4
survival *see* cats; chick, survival; foxes
*Sylviornis neocaledoniae* 12

tarsus length 20, 24, 26
temperature-dependent sex ratios 59, 64–5
temperature sensor 38

underground nesting, adaptations to 57–61
ultraviolet (UV) reflection
legs and beak 77
wattle 19, 87

vocalisations 21–7
Vogel, U. 74
wattle, *see* Australian Brush-turkey, wattle
weather prophets 47–8
Wheeler, R. 46, 74, 79, 98

## OTHER TITLES IN THE AUSTRALIAN NATURAL HISTORY SERIES: